T0239442

SpringerBriefs in Electrical and Computer Engineering

Series Editors

Woon-Seng Gan, School of Electrical and Electronic Engineering, Nanyang Technological University, Singapore, Singapore

C.-C. Jay Kuo, University of Southern California, Los Angeles, CA, USA

Thomas Fang Zheng, Research Institute of Information Technology, Tsinghua University, Beijing, China

Mauro Barni, Department of Information Engineering and Mathematics, University of Siena, Siena, Italy

SpringerBriefs present concise summaries of cutting-edge research and practical applications across a wide spectrum of fields. Featuring compact volumes of 50 to 125 pages, the series covers a range of content from professional to academic. Typical topics might include: timely report of state-of-the art analytical techniques, a bridge between new research results, as published in journal articles, and a contextual literature review, a snapshot of a hot or emerging topic, an in-depth case study or clinical example and a presentation of core concepts that students must understand in order to make independent contributions.

More information about this series at http://www.springer.com/series/10059

Ahmed A. Zaki Diab ·
Abo-Hashima M. Al-Sayed ·
Hossam Hefnawy Abbas Mohammed ·
Yehia Sayed Mohammed

Development of Adaptive Speed Observers for Induction Machine System Stabilization

 Springer

Ahmed A. Zaki Diab
Electrical Engineering Department
Faculty of Engineering
Minia University
Minia, Egypt

Department of Electrical
and Electronic Engineering
Kyushu University
Fukuoka, Japan

Hossam Hefnawy Abbas Mohammed
Electricity and Water Authority
Manama, Kingdom of Bahrain

Electrical Engineering Department
Faculty of Engineering
Minia University
Minia, Egypt

Abo-Hashima M. Al-Sayed
Electrical Engineering Department
Faculty of Engineering
Minia University
Minia, Egypt

Yehia Sayed Mohammed
Electrical Engineering Department
Faculty of Engineering
Minia University
Minia, Egypt

ISSN 2191-8112   ISSN 2191-8120   (electronic)
SpringerBriefs in Electrical and Computer Engineering
ISBN 978-981-15-2297-0   ISBN 978-981-15-2298-7   (eBook)
https://doi.org/10.1007/978-981-15-2298-7

This Springer imprint is published by the registered company Springer Nature Singapore Pte Ltd.
The registered company address is: 152 Beach Road, #21-01/04 Gateway East, Singapore 189721, Singapore

About This Book

The book presents a comprehensive design of the adaptive state observers for induction machines drives. Experimental and simulation results have been introduced to demonstrate the effectiveness of the presented control schemes. The book objectives are summarized as follows:

1. Development of the mathematical model of an adaptive state observer based on the induction motor model taking the parameter variations into consideration to achieve high performance for sensorless induction motor drives.
2. Derive an expression for a modified gain rotor flux observer with a parameter adaptive scheme based on the machine model. This observer is used to estimate the motor speed accurately and improve the stability and performance of sensorless vector control induction motor drives.
3. Implementation of a sensorless vector control scheme to control the VSI fed from PV array for the induction motor water pumping system. An adaptive full order state observer utilized to estimate the motor speed for increasing the overall system reliability. The proposed control scheme has been simulated under different operating condition of varying irradiation and temperature.
4. Designing of a robust vector-controlled induction motor drives based on H_infinty theory for speed estimation using MRAS to achieve specific design requirement. The effectiveness of this controller is assessed by comparing the calculated results with the PI controller.

Introduction

The speed sensorless vector control techniques of induction motor drives are used in high performance applications. These techniques generally require an accurate determination of the machine parameters, rotor flux and motor speed. It uses a full order-type adaptive rotor flux observer that takes parameters variation into account to achieve high steady state performance without spoiling the dynamic response.

In this book, an adaptive state observer is derived based on the induction machine model. A modified gain rotor flux observer with a parameter adaptive scheme has been proposed for sensorless vector control induction motor drives. The optimal value of the observer gain has been proved by minimizing the error between the measured and estimated states. Also, the stability of the proposed observer with the parameter adaptation scheme is proved by the Lyapunov's theorem.

The application of sensorless vector control drives to control a photovoltaic (PV) motor pumping system has been presented in this book. The principle of vector control has been applied to control the single stage of voltage source inverter (VSI) feeding the three-phase induction motor. The main objective of this work is to design and analyze a single stage maximum power point tracking (MPPT) from PV module and eliminate the speed encoder. Elimination of the speed encoder aims to increase the reliability of the PV motor pumping system. Therefore, a full order adaptive state observer has been designed to estimate the rotor speed of the motor pumping system. Moreover, the incremental conductance method is used for achieving MPPT of the PV system. The control scheme with full order adaptive state observer has been investigated under different operating conditions of varying natures of solar radiation and air temperature. The simulation results show that the response of the PV motor pumping system with the adaptive speed observer has a good dynamic performance under different operating conditions.

In order to improve the dynamic response and to overcome the problems of the classical speed controller (such as overshoot, long settling time and oscillations of motor speed and torque), a robust controller design based on the H_∞ theory for high performance sensorless induction motor drives is implemented. The proposed controller is robust against system parameter variations and achieves good dynamic performance. In addition, it rejects disturbances well and can minimize system noise. The H_∞ controller design has a standard form that emphasizes the selection of the weighting functions that achieve the robustness and performance goals of motor drives in a wide range of operating conditions. The model reference adaptive system (MRAS) is used to estimate the motor speed based on the measurement of stator voltages and currents. In this work, the stability of the estimator scheme is discussed and proved based on Popov's criterion. To investigate the effectiveness of the proposed control scheme at different operating conditions (such as a sudden change of the speed command/load torque disturbance), its performance is compared with those of the classical control one. Experimental and simulation results demonstrate that the presented control scheme with the H_∞ controller and MRAS speed estimator has an accurate estimated motor speed and a good dynamic performance.

Contents

About the Authors

Ahmed A. Zaki Diab, Ph.D. received B.Sc., and M.Sc. in Electrical Engineering from Minia University, Egypt, in 2006 and 2009, respectively. In 2015, he received Ph.D. from Electric Drives and Industry Automation Department, Faculty of Mechatronics and Automation at Novosibirsk State Technical University, Novosibirsk, Russia. He had obtained a postdoctoral Fellowship at the National research university "MPEI," Moscow Power Engineering Institute, Moscow, Russia, from September 2017 to March 2018. Since 2007, he has been with the Department of Electrical Engineering, Faculty of Engineering, Minia University, Egypt as a teaching assistant, a lecturer assistant, and since 2015, as an assistant professor. Currently, he is a visitor researcher (postdoctoral) at Green Power Electronics Circuits Laboratory, Kyushu University, Japan (awarded the MIF Research Fellowship 2019, Japan). He is certified as a Siemens Engineer and Trainer in several fields of automation and process control systems, and he is a supervisor of the Automatic Control and Traction Laboratory, Faculty of Engineering, Minia University, Egypt. His present research interests include Renewable Energy Systems, Power Electronics, and Machines Drives.

Abou-Hashema M. Al-Sayed, Ph.D. received his B.Sc., and M.Sc. in Electrical Engineering from Minia University, Minia, Egypt, in 1994 and 1998, respectively. He was a Ph.D. student in the Institute of Electrical Power Systems and Protection, Faculty of Electrical Engineering, Dresden University of Technology, Dresden, Germany, from 2000 to 2002. He received his Ph.D. in Electrical Power from the Faculty of Engineering, Minia University, Egypt, in 2002, according to a channel system program, which means a Scientific Co-operation between the Dresden University of Technology, Germany and Minia University, Egypt. Since 1994, he has been with the Department of Electrical Engineering, Faculty of Engineering, Minia University, as a teaching assistant, a lecturer assistant, an assistant professor, and associate professor. He was a visiting researcher at Kyushu University, Japan, from 2008 to 2009. He is the head of Mechatronics and Industrial Robotics Program, Faculty of Engineering, Minia University, from 2011 till now. His research interests include Protection Systems, Renewable Energy, and Power Systems.

Hossam Hefnawy Abbas Mohammed, Ph.D. received his B.Sc., M.Sc., and Ph.D. in Electrical Engineering from Department of Electrical Engineering, Faculty of Engineering, Minia University, Egypt, in 2004, 2015 and 2019, respectively. He has joined Middle Egypt Electricity Distribution Company as an electrical engineer in 2004. Also, he had worked in Telecom Egypt Co. as a head of Control Center Devices Operation and Maintenance Department in 2008. He is currently working in Electricity and Water Authority in Kingdom of Bahrain. He had more than fifteen years of practical working experience. His research interests cover power electronics, electric machines control, renewable energy, modeling and simulation of power converters, computer algebra systems, power quality and harmonics analysis, advanced control theory, and DSP-based control applications.

Yahia Sayed Mohammed, Ph.D. received his B.Sc., M.Sc., and Ph.D. in Electrical Engineering from Assiut University, Assiut, Egypt, in 1983, 1989, and 1998, respectively. He is a professor and the head of Electrical Engineering Department, Faculty of Engineering, Minia University, Minia, Egypt. His research interests include Machine Drives and Renewable Energy.

Abbreviations

AC	Alternating current
AINC	Adaptive incremental conductance
AM	Air mass
CCS	Code Composer Studio
CRPWM	Current regulated pulse width modulation
DC	Direct current
DSP	Digital signal processor
FOC	Field-oriented control
IFO	Indirect field oriented
IM	Induction motor
IMD	Induction motor drive
LPV	Linear parameter varying
MPP	Maximum power point
MPPT	Maximum power point tracking
MRAS	Model reference adaptive system
PI	Proportional Integral
PV	Photovoltaic
PWM	Pulse width modulation
SR	Solar radiation
STC	Standard test conditions at (25°C and 1000 W/m^2)
VSI	Voltage source inverter

Symbols

$e_{m\alpha}, e_{m\beta}$	Stationary counter EMF components
f_d	Damping coefficient
I_{sc}	Short-circuit current
$i_{\alpha s}, i_{\beta s}$	Stationary axis stator current components
$\hat{i}_s$	Estimated stator current
i_s	Stator current
I, J	Unit matrix and skew symmetric matrix
J_m	Moment of inertia
L_r	Rotor self-leakage inductance
L_s	Stator self-leakage inductance
L_s	Stator self-inductance
L_r	Rotor self-inductance
N_{ser}	the number of modules connected in series
N_{par}	the number of modules connected in parallels
P	Number of pole pairs
$p = \frac{d}{dt}$	Differential operator
q	The electron charge (1.60217646e-19 C)
R_s	Stator resistances
R_r	Rotor resistances
T_l	Load torques
T_e	Electromagnetic torques
T_r	Rotor time constant
V_{oc}	Open-circuit voltage
V_{T1} and V_{T2}	Thermal junction voltage of PV model
V_{ds}, V_{qs}	d-q axis stator voltage components
$V_{\alpha s}, V_{\beta s}$	Stationary axis stator voltage components
V_s	Stator voltage
σ	Leakage coefficient
ψ_r	Rotor flux vector
ψ_{dr}, ψ_{qr}	d-q axis rotor flux components

$\psi_{\alpha r}, \psi_{\beta r}$	Stationary axis rotor flux components
$\hat{\psi}_r$	Estimated rotor flux
ω_r	Rotor angular speed
$\hat{\omega}_r$	Estimated rotor angular speed
$\wedge$	Signifies the estimated value
*	Denotes the reference or command value

List of Figures

Chapter 1
Introduction and Background of Induction Machine System Stabilization

Abstract The development of high-performance induction motor drives for general industry applications and production automation has received wide spread research interests. Many schemes have been proposed for the control of induction motor drives, among which the field-oriented control or vector control, being accepted as one of the most effective methods.

Keywords Induction motor drives · Field-oriented control · Speed observers · H_infinity · Experimental implementation

The development of high-performance induction motor drives for general industry applications and production automation has received wide spread research interests. Many schemes have been proposed for the control of induction motor drives, among which the field oriented control [1, 2] or vector control, being accepted as one of the most effective methods. This is achieved by decoupling the stator current (flux-and torque-producing) components in order to obtain a decoupled control of the flux and electromagnetic torque exactly as a separately excited dc motor.

This is implemented by controlling the so-called "field angle" defining the angular position of the rotor flux with respect to the stator fixed frame.

The field angle is measured or calculated from an estimator or observer in the so-called direct field-oriented control scheme [1]. On the other hand, the field angle is obtained as equal to the sum of an estimated slip-angle and the rotor position angle in the so-called indirect field-oriented control scheme [3]. However, all the field orientation methods suffer from some specific practical problems. In the indirect methods, the slip angle depends generally on the value of rotor time-constant which varies with temperature, saturation and load. Direct methods, which have a minor dependency on rotor parameters, suffer from measurement instability of the flux vector at low speeds at a time where Hall sensors are rarely used in industrial applications. Only estimation or observation of the flux vector is possible with direct method and is generally done starting from the machine-terminal variables and using a machine model with related parameters.

In vector control schemes, the detection of rotor speed is necessary for calculating the field angle and adjustment of the speed in a feedback-loop. However, the speed sensor spoils the ruggedness and simplicity of induction motors, so that the elimination of this sensor has been one of the important requirements in vector-control schemes. Many investigations about speed sensorless vector-control were reported [4–17], but little attention was paid to the stability of adaptive speed observer with parameter adaptive scheme on motor performance. This motivates the author to develop this speed observer by modifying the observer gain to improve the system stability in a wide range of operating conditions.

In sensorless vector control scheme, the stator circuit resistance is changes with motor temperature which leads to instability of the estimated speed especially at low speeds. In order to compensate the effect of stator circuit resistance which is changed with temperature and frequency, the speed and stator resistance are estimated simultaneously. These estimations are based on adaptive state observer for sensorless vector-controlled induction motor scheme. One of the important applications of the sensorless control scheme is the implementation for controlling the VSI fed from PV array for motor pumping system at different irradiation levels and ambient temperatures.

In sensorless vector control methods, the classical proportional plus integral (PI) controller was used in speed control schemes because of its simplicity and stability. The performance of the PI controller often gives poor dynamic responses for changes in the load torque and moment of inertia. To overcome these drawbacks of the PI controller, a robust controller based on H_infinity theory is proposed to achieve specific design requirements for sensorless vector-controlled motors with using an adaptive speed observer.

1.1 Research Objectives

In view of the foregoing brief discussion, the book objectives are summarized as follows:

1. Development of the mathematical model of an adaptive state observer based on induction motor model taking the parameter variations into consideration to achieve high performance for sensorless induction motor drives.
2. Derive an expression for a modified gain rotor flux observer with a parameter adaptive scheme based on machine model. This observer is used to estimate the motor speed accurately and improve the stability and performance of sensorless vector control induction motor drives.
3. Implementation of a sensorless vector control scheme to control the VSI fed from PV array for induction motor water pumping system. An adaptive full order state observer utilized to estimate the motor speed for increasing the overall system reliability. The proposed control scheme has been simulated under different operating condition of varying irradiation and temperature.

4. Designing of a robust vector-controlled induction motor drives based on H_infinty theory for speed estimation using MRAS to achieve specific design requirement. The effectiveness of this controller is assessed by comparing the calculated results with the PI controller.

1.2 Organization of the Book

The present book is organized in six chapters and three appendices. Apart from the introduction and conclusions chapters, four chapters form the body of the book.

Chapter 2 is devoted to a general review of literature pertinent to sensorless induction motor drives applications and the different methods of motor speed estimation performance.

In Chap. 3, a development of an adaptive state observer based on mathematical model of an induction motor as influenced by motor parameter variations is proposed. In this chapter, a modified gain rotor flux observer with a parameter adaptive scheme and the stability of the proposed observer is proved by the Lyapunov's theorem. This observer is implemented to estimate the motor speed accurately for sensorless induction motor drive application. Advanced MRAS based on machine model using the voltages and currents measurements to estimate the motor speed for high performance sensorless induction motor drives is presented. Also, system stability has been derived in this chapter.

In Chap. 4, a method of speed and stator resistance estimations based on adaptive state observer for sensorless vector-controlled induction motor scheme is proposed. This control scheme is used to control the VSI fed from PV array for water pumping system. Also, the presented work in this chapter is to design and analyze the single stage MPPT system. A digital simulation of the proposed system under different operating condition of varying natures of solar irradiation and air temperature is presented. The accuracy and correctness of the simulation algorithm validate that the proposed system has a high dynamic performance along with different operating conditions.

In Chap. 5, a robust speed controller design for a sensorless vector-controlled induction motor drive system based on H_infinity theory is presented to overcome the problems of the classical controller (PI controller). The motor speed is estimated based on stabilizing MRAS. The structure of MRAS is presented in this chapter based on machine model using stator voltages and currents measurements. The proposed controller is designed to select the weighting functions for achieving the robustness and performance goals of motor drives in a wide range of operating conditions. The effectiveness of the proposed system is investigated in this chapter by comparing the calculated motor speed, torque, stator phase currents, stator current components and rotor flux due to step change of the speed command/load torque with those obtained using classical PI controller. Experimental results are also included in this chapter to demonstrate the validity of the proposed system.

Chapter 6 summarizes the main conclusions drawn from the research work reported in the book along with recommendations proposed for further research in the area of adaptive speed observers for induction machine system stabilization.

In addition to these chapters, a quite useful list of references pertinent to the topics treated in the book is given.

For related details, the book is ended with three appendices summarized as follows:

Appendix A reports the specifications data of induction machine parameter and the parameter of PV water pumping system used for sensorless vector control scheme which control the VSI fed from PV array.

Appendix B presents the parameters and data specifications of the test motor used for sensorless high performance induction motor drives with robust controller.

Appendix C includes a description of the experimental set-up and technique used for (Texas Instruments) the data specifications of the microcontroller kit of digital signal processor TMS28F035 series.

References

1. Blaschke F (1972) The principle of field orientation as applied to the new transvector closed loop control system for rotating machines. Siemens Rev 39(5):213–220
2. Gabriel P, Leonhard W, Nordby CJ (1980) Field-oriented control of a standard AC motor using microprocessor. IEEE Trans Ind Appl 369(2):182–192
3. Sugimoto H, Ohno E (1984) Decoupled-control and characteristics of induction motor driven by variable voltage/frequency source. Inst Elec Eng Japan Trans 104(11):123–129
4. Radu B, Paolo G, Gian-Mario P (2008) Sensorless direct field-oriented control of three-phase induction motor drives for low-cost applications. IEEE Trans Ind Appl 44(2):475–481
5. Ichiroc M, Youichi O (1991) Improvement of robustness on speed sensorless vector control of induction motor. EPE Firenze 4:660–665
6. Ohtanti T, Takada N, Tanaka K (1992) Vector control of induction motor without shaft encoder. IEEE Trans Ind Appl 28(1):157–164
7. Kubota H, Matuse K, Nakano T (1993) DSP—based adaptive speed flux observer of induction motor. IEEE Trans Ind Appl 29(2):344–348
8. Tajima H, Hori Y (1993) Speed sensorless field orientation control of induction machine. IEEE Trans Ind Appl 2(1):175–180
9. Bonanno GJ, Zhen L, Xu L (1995) A direct field oriented induction machine drive with robust flux estimator for position sensorless control. In: Conference record IEEE-IAS annual conference, pp 166–173
10. Benlaloui I, Drid S, Chrifi-Alaoui L, Ouriagli M (2015) Implementation of a new MRAS speed sensorless vector control of induction machine. IEEE Trans Energy Convers 30(2):588–595
11. Yang G, Chin TH (1993) Adaptive speed identification scheme for a vector controlled speed sensorless inverter-induction motor drive. IEEE Trans Ind Appl 29(4):1054–1062
12. Kanmachi T, Takahashi I (1993) Sensorless speed control of an induction motor with no influence of secondary resistance variation. In: Conference record of the IEEE-IAS, pp 408–413
13. Kim YR, Sul SK, Park MH (1994) Speed sensorless vector control of induction motor using extended Kalman filter. IEEE Trans Ind Appl 30(5):1225–1233
14. Hurst KD, Habetler TG, Griva G, Profumo F (1994) Speed sensorless field oriented control of induction machines using current harmonic spectral estimation. In: Conference record of the IEEE-IAS, vol 1, pp 601–607

15. Ilas C, Bettini A, Ferraris L, Griva G, Profumo F (1994) Comparison of different schemes without shaft sensors for field oriented control drives. In: Proceedings of IEEE-IECON, vol 3, pp 1579–1588
16. Bose BK, Simoes MG, Crecelius DR, Rajashekara K, Martin R (1995) Speed sensorless hybrid vector controlled induction motor drive. In: Conference record of the IEEE-IAS, vol 1, pp 137–143
17. Wook Kim H, Ki Sul S (1996) A new motor speed estimator using Kalman filter in low-speed range. IEEE Trans Ind Electron 43(4):498–504

Chapter 2
Literature Review of Induction Motor Drives

Abstract The development of effective induction motor drives for various applications in industry has received intensive effort for many researchers.

Keywords Induction motor drives · Field-oriented control · H_infinity · Speed observers · Rotor flux estimators

2.1 Introduction

The development of effective induction motor drives for various industrial applications has received intensive effort for many researchers. Several methods have been developed to control induction motor drives such as scalar control, field-oriented control and direct torque control, among which field-oriented control is one of the most successful and effective methods [1–5]. In field orientation, with respect to using the two-axis synchronously rotating frame, the phase current of the stator is represented by two component parts; the field current part and the torque-producing current part. When the component of the field current is adjusted constantly, the electromagnetic torque of the controlled motor is linearly proportional to the torque-producing components, which is comparable to the control of a separately excited DC motor. The torque and flux are considered as input commands for a field-oriented controlled induction motor drive, while the three-phase stator reference currents after a coordinated transformation of the two-axis currents are considered as the output commands. To achieve the decoupling control between the torque and flux currents components, the three-phase currents of the induction motor are controlled so that they follow their reference current commands through the use of current-regulated pulse-width-modulated (CRPWM) inverters [2–8]. Moreover, the controls of the rotor magnetic flux level and the electromagnetic torque are entirely decoupled using an additional outer feedback speed loop. Therefore, the control scheme has two loops; the inner loop of the decoupling the currents components of flux and torque, and the outer control loop which controls the rotor speed and produces the reference electromagnetic torque. Based on that, the control of an induction motor drive can be considered as a multi feedback-loop control problem consisting of current control and speed-control loops.

A. A. Z. Diab et al., *Development of Adaptive Speed Observers for Induction Machine System Stabilization*, SpringerBriefs in Electrical and Computer Engineering, https://doi.org/10.1007/978-981-15-2298-7_2

2.2 Rotor Flux Estimator for Sensorless Induction Motor

The rotor flux vector in Fig. 2.1 is estimated by integrating the back e.m.f calculated from stator currents and voltages [9]. This method suffers from the poor accuracy of

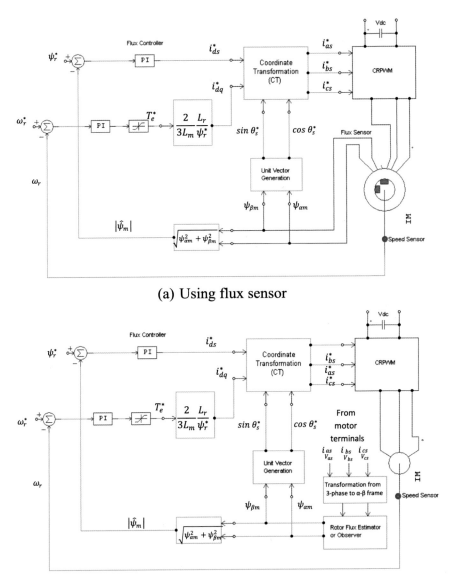

(a) Using flux sensor

(b) Using rotor flux estimator or observer

Fig. 2.1 Block diagram of direct vector control induction motor drive

flux estimation in the low-speed region arising from the integration problem and the change of stator resistance with temperature.

On the other hand, the rotor flux vector is estimated from stator currents and the rotor speed with no need for integration, but on the basis of the rotor circuit equations of the induction motor in the α-β stationary reference frame [10]. The rotor equation in the β-axis stationary reference frame can be obtained by putting $\omega_s = 0$ in Eq. (2.5) and is expressed as:

$$\frac{d\psi_{\beta r}}{dt} + i_{\beta r} R_r - \omega_r \psi_{\alpha r} = 0 \tag{2.1}$$

In the above equation, adding the term $\frac{R_r L_m}{L_r} i_{\beta s}$ on both sides, will give

$$\frac{d\psi_{\beta r}}{dt} + \frac{R_r}{L_r}(L_r i_{\beta r} + L_m i_{\beta s}) - \omega_r \psi_{\alpha r} = \frac{R_r L_m}{L_r} i_{\beta s} \tag{2.2}$$

The α-β components of rotor flux are expressed in terms of the α-β components of stator and rotor currents as [10]:

$$\psi_{\alpha r} = L_r i_{\alpha r} + L_m i_{\alpha s} \tag{2.3}$$

$$\psi_{\beta r} = L_r i_{\beta r} + L_m i_{\beta s} \tag{2.4}$$

Substituting Eq. (2.4) into Eq. (2.3), the following equation is obtained after simplification:

$$\frac{d\psi_{\beta r}}{dt} = \frac{L_m}{T_r} i_{\beta s} - \frac{\psi_{\beta r}}{T_r} + \omega_r \psi_{\alpha r} \tag{2.5}$$

Similarly, the equation from the α-axis stationary frame can be derived as:

$$\frac{d\psi_{\alpha r}}{dt} = \frac{L_m}{T_r} i_{\alpha s} - \frac{\psi_{\alpha r}}{T_r} + \omega_r \psi_{\beta r} \tag{2.6}$$

The rotor flux is obtained as a function of rotor speed and stator currents by using Eqs. (2.5) and (2.6) and the simulation diagram for estimation is shown in Fig. (2.2). Therefore, the rotor flux vector can be estimated even at standstill condition. The major weakness of this method is the sensitivity of estimation to the change of rotor parameters arising from the magnetic saturation and change of winding temperature.

In field-orientation control schemes, the efficiency of the induction motor was improved when the rotor flux level was changed according to the load [11–14]. In order to obtain high efficiency of high power induction motors without spoiling the good speed and torque response, a method was proposed [11–14] to control rapidly the rotor flux level with deadbeat-controller.

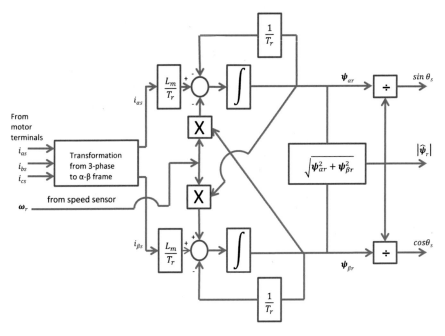

Fig. 2.2   Rotor flux estimation from speed and rotor currents

2.3   Rotor Flux Observer for Sensorless Induction Motor

Recently, rotor flux observers have been proposed for rotor flux calculation [15–17]. The flux predicted from the rotor flux observer is less influenced by the winding (stator and rotor) resistance-variations with temperature, when compared with the rotor flux estimator, especially in the low-speed range. A rotor flux observer with parameter adaptive scheme was presented [18–24] to improve the flux prediction in the low-speed range. The parameters identified adaptively in this observer are stator and rotor resistances which are dependent upon the motor temperature.

The induction motor can be described by state equations in the α-β axis stationary reference frame [25]. The state variables are the stator current $i_s = \begin{bmatrix} i_{\alpha r} & i_{\beta r} \end{bmatrix}^T$ and the rotor flux as: $\psi_r = \begin{bmatrix} \psi_{\alpha r} & \psi_{\beta r} \end{bmatrix}^T$. The input variable is the stator voltage.

$$V_s = \begin{bmatrix} V_{\alpha s} & V_{\beta s} \end{bmatrix}^T$$

A rotor flux observer was designed by using Gopinath's reduced order observer theory [26]. The rotor flux observer equation takes the form:

$$\frac{d\widehat{\psi}_r}{dt} = A_{22}\widehat{\psi}_r + A_{21}i_s + G\left(\frac{di_s}{dt} - A_{12}\widehat{\psi}_r - A_{11}i_s - B_1 V_s\right) \qquad (2.7)$$

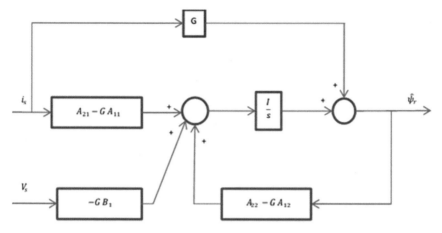

Fig. 2.3 Configuration of the rotor flux observer

where A_{11}, A_{12}, A_{21} and A_{22} are function of motor parameters and speed. And G is the observer gain [27].

The unit vector was estimated from the two component of rotor flux ($\psi_{\alpha r}$ $\psi_{\beta r}$) as:

$$\cos\theta_s = \frac{\psi_{\alpha r}}{\widehat{\psi}_r} \quad \text{and} \quad \sin\theta_s = \frac{\psi_{\beta r}}{\widehat{\psi}_r}$$

where $\widehat{\psi}_r = \sqrt{\psi_{\alpha r}^2 + \psi_{\beta r}^2}$

The configuration of the rotor flux observer is shown in Fig. 2.3.

2.4 Effect of Speed Estimation Methodology on Sensorless Induction Motors

In flux regulation by current and slip frequency control schemes, the detection of rotor speed is necessary for establishing the outer feedback-loop of speed. However, in field orientation control schemes, the detection of rotor speed is necessary for calculating the field angle and establishing the outer feedback-loop of speed.

Recently, the elimination of speed sensor has been one of the important requirements in vector-control schemes because the speed sensor spoils the ruggedness and simplicity of induction motors control systems.

A modified rotor flux estimator has been proposed [28] for direct vector-controlled induction motor drives without speed sensor. The estimated rotor flux is not only due to the integration of the back-e.m.f but also due to the magnetizing current.

This improves the accuracy of the estimator at low speed because the magnetizing current contributes heavily to the flux estimation in comparison with the back-e.m.f integration.

An estimator for the motor speed was proposed [29] and applied to the flux observer based field orientation control system. The speed was estimated from the difference between two rotor flux estimators, one was based on stator equations (reference model) and the other on the rotor equations (adjustable model). This is referred to as a model reference adaptive system (MRAS).

A speed estimation scheme was proposed [30] based on a modified MRAS to implement a speed sensorless indirect field-oriented control of an induction motor. In this scheme, the reference model was modified by eliminating the need for evaluating the stator leakage inductance without a pure integration. This results in good performance over a wide speed range. To obtain a more accurate estimation of speed, a simple on-line rotor time-constant estimation has been incorporated.

A model reference adaptive system (MRAS) for a speed sensorless vector-controlled induction-motor drive was presented [31]. The rotor speed was estimated with a full-order adaptive-flux observer and was used as the feedback signal for the vector control and the speed control. In order to estimate low speed values accurately, the motor stator resistance was identified at the same time to correct the mismatched resistance values used in the observer.

Many investigations about speed sensorless vector-control were reported [32–41], but no attention was paid to the stability of adaptive speed observer with considering parameter variation on motor performance. This motivates the author to develop the optimal observer gain of this speed observer to achieve the system stability.

2.5 Adaptive Speed Observer for Sensorless Vector Control for Motor Pumping System Applications

Water pumping system is an energy intensive process. However, the conventional pumping systems are powered from fossil fuels. The costs of fossil fuels are increasing and cause additional harming to the environment with high pollution and greenhouse gas emissions [42–47]. In recent years, the solar energy is used to provide electricity for water pumping systems in the isolated remote area in many countries [42]. Egypt is a promising country for generating the electric energy from the sun because it has the highest number of sunshine hours all year round with higher irradiation levels [43].

The water pumping systems based on the solar energy is essential for development projects in remote areas. Moreover, the PV pumping systems will help in desert development and supporting the Egyptian economy. However, with low efficiency of the solar cell, the MPPT techniques are applied to improve the characteristic of the PVs. Moreover, MPPT leads to make the PV systems work efficiently and effectively in different climatic conditions. The most popular type of electrical motors, which are used for pumping systems are DC motors and Induction motors. However, DC motors have many drawbacks originated from frequently maintenance problems related to the presence of the commutator and brushes. So, the applications of the pumping systems based on induction motors are the attractive proposal with high reliability, speed capability, robustness and maintenance-free operations [48].

Furthermore, the application of the field-oriented control theory with induction motor results in improving the performance of the control scheme for the high-performance applications [44].

The elimination of the rotor speed encoder and estimation of the rotor speed in sensorless IM drives results a high reliability. Thus, this book presents the PV pumping system based on PV array, DC link capacitor, a voltage source inverter, three phase induction motor, maximum power point tracking (MPPT) technique, Vector control method is used for controlling the induction machine. Moreover, the full order observer has been applied for estimating the rotor speed.

PV array is built up using series and parallel connections of PV panels for matching the required power, voltage and current rating of the motor. Here a voltage source inverter (VSI) directly converts DC power to AC power, without need to DC-DC converter and save its cost. It in turn reduces the whole system size, as there is no need for inductor. Also in PV pump storage system, when sunlight is available, solar energy is stored in the shape of potential energy in water tank storage, and discharged in the period of high demand for electric energy instead of using battery storage systems, which are heavy, more expensive and have low lifetime equal to one fifth of that for the PV system. So the large battery banks are not advisable in this study depending on the previously mentioned reasons [45]. Vector control is used for the induction machine control, which improves its performance over the scaler variable frequency drive [29, 33–35]. Operating point, which is defined as the point of intersection between the voltage- current characteristics of the PV panel and the characteristics of the pump-motor set may be far from the MPP of the PV system that leads to a significant loss of the available power generated by the PV modules [49, 50]. In the control system, the maximum power point tracking unit is used to optimize the operating point for power extraction by decreasing and increasing the value of reference speed fed to the vector control to match the PV generator with the optimum motor-pump operating point.

Recently, many approaches are presented for sensorless control of the IM, for increasing the reliability of the drive. Numerous algorithms are presented for improving the estimation accuracy of the induction machines sensorless drives [51–54]. However, the stator resistance value has an important effect on the performance of the sensorless drives. Moreover, the value of the stator resistance should be accurate estimated with an acceptable precision to insure an accurate estimation of the lower values of the rotor speed [51–54]. This motivates the author to estimate the stator resistance in parallel with rotor speed to improve the performance and accuracy of the full order state observer.

2.6 Effect of Parameter Variation of the Speed Controllers on Sensorless Induction Motors

2.6.1 Speed Controller Based on Classical PI Controller

In vector control schemes, the classical Proportional-Integral (PI) controller is frequently used in a speed-control loop due to its simplicity and stability. The parameters of the PI controller are designed through trial-and-error [20–22]. However, PI controllers often yield poor dynamic responses to changes in the load torque and moment of inertia. To overcome this problem of classical PI controllers and to improve the dynamic performance, various approaches have been proposed in Refs. [23, 24]. The classical two-degree-of-freedom controller (phase lead compensator and PI controller) [23] was used for indirect vector control of an induction motor drive. However, the parameters of this controller are still obtained through trial-and-error to reach a satisfactory performance level.

2.6.2 Speed Controller Based on Fuzzy Logic Controller

In Ref. [25], the authors presented a control scheme for the induction motors drives based on fuzzy logic. The proposed control scheme has been applied to improve the overall performance of an induction motor drive system. This controller does not require a system model and it is insensitive to external load torque disturbances and information error. On the other hand, the presented control scheme suffers from drawbacks such as large oscillations in transient operation. Moreover, the control system requires an optical encoder to measure the motor speed [25].

2.6.3 Speed Controller Based on Linear Quadratic Gaussian Method

A linear quadratic Gaussian controller was applied in Refs. [55, 56] to regulate motor speed and improve the motor's dynamic performance. The merits of this controller are as follows: fast response, robustness and the ability to operate with available noise data. However, this controller's drawbacks are that it needs an accurate system model, does not guarantee a stability margin and requires more computation.

2.6.4   Speed Controller Based on H_Infinity Control Theory

Recently, H_infinity (H_∞) control theory has been widely implemented for its robustness against model uncertainty perturbations, external disturbances and noise. Some applications of this technique in different systems, such as permanent magnet DC motors [57], switching converters [58] and synchronous motors [59], have been reported. Moreover, researchers have worked to apply the H_∞ controllers in the induction machines drives.

In Ref. [60], a control scheme based on the H_∞ is presented for control the speed of the induction motors. However, the control system is validated only through the simulation results. Additionally, the speed sensor which used to measure the rotor speed reduces the control system reliability and also increases its cost. In Ref. [61], a vector control scheme for the induction machines based on H_∞ has been designed and experimentally validated. However, the authors used a sensor to measure the rotor speed. Additionally, a comparison between the performances of the sensor vector-control scheme of induction motor based on the PI controller is presented. The results show the superiority of the H_∞ control scheme rather than the PI controller. The main drawbacks of the control system of Ref. [61] were that speed sensor data simulation verification results were not included. Another control scheme based on the H_∞ has been presented in Ref. [62].

The introduced control scheme in Ref. [62] has many drawbacks such as the need for a speed encoder, and the control law which is based on the linear parameter varying (LPV) should be updated online which increases the cost of implementation. However, validation of the control scheme has been carried out based on only a simulation using the MATLAB/Simulink (2014a, MathWorks, Natick, MA, USA) package. The authors of this paper recommended future work to eliminate the speed sensor and to minimize the implementation time. An interesting research work about the application of induction machines drives has been presented in Ref. [63].

The control scheme is applied for Electric Trains application. The control system suffers from reliability reduction because of presence of the speed sensor and also the increasing of the implementation time because of the time which is needed to reach the solution of the Riccati equation. From the previous discussion, further research work is required to enhance the dynamic performance of the induction motor drives with the application of the H_∞ control theory. Moreover, the application of sensorless algorithms with the H_∞ based induction machine drives is an essential research point. Furthermore, the experimental implementation of the induction machines drives based on H_∞ is required for greater validation of the control scheme. Additionally, the major aspect of an H_∞ control is to synthesize a feedback law that forces the closed-loop system to satisfy a prescribed H_∞ norm constraint. This aspect achieves the desired stability and tracking requirements. This motivates the author to use a robust controller for sensorless vector-controlled induction motor drives with using MRAS to estimate the motor speed.

References

1. Vas P (1990) Vector control of AC machines, vol 22. Oxford University Press, Oxford, UK, pp 122–215
2. Krishnan R, Doran FC (1987) Study of parameter sensitivity in high performance inverter fed induction motor drive systems. IEEE Trans Ind Appl IA-23(4):623–635
3. Vas P (1993) Parameter estimation, condition monitoring, and diagnosis of electrical machines, vol 27. Oxford Science Publications, Oxford, UK
4. Toliyat HA, Levi E, Raina M (2003) A review of RFO induction motor parameter estimation techniques. IEEE Trans Energy Convers 18(2):271–283
5. Minami K, Veles-Reyes M, Elten D, Verghese GC, Filbert D (1991) Multi-stage speed and parameter estimation for induction machines. In: Proceedings of the record 22nd annual IEEE power electronics specialists conference (IEEE PESC'91), Cambridge, MA, USA, 24–27 June 1991, pp 596–604
6. Roncero-Sánchez PL, García-Cerrada A, Feliú V (2007) Rotor-resistance estimation for induction machines with indirect-field orientation. Control Eng Pract 15(9):1119–1133
7. Karayaka HB, Marwali MN, Keyhani A (1997) Induction machines parameter tracking from test data via PWM inverters. In: Proceedings of the record of the 1997 IEEE industry applications conference thirty-second IAS annual meeting, New Orleans, LA, USA, 5–9 Oct 1997, pp 227–233
8. Godoy M, Bose BK, Spiegel RJ (1996) Design and performance evaluation of a fuzzy-logic-based variable-speed wind generation system. IEEE Trans Energy Convers 1:349–356
9. Gabriel P, Leonhard W, Nordby CJ (1980) Field-oriented control of a standard AC motor using microprocessor. IEEE Trans Ind Appl IA-16(2):186–192
10. Bose BK (2001) Modern power electronics and AC drives. Prentice Hall PTR, NJ
11. Kim HG, Sul SK, Park MH (1984) Optimal efficiency drive of current source inverter fed induction motor by flux control. IEEE Trans Ind Appl IA-20(6):1453–1459
12. Matsuse K, Kubota H (1987) Deadbeat response of flux control of induction motor. EPE 2:895–898
13. Matsuse K, Kubota H (1987) Digital control scheme of rotor flux in induction motor with the deadbeat response. In: Conference record, IEEE-IAS, pp 222–226
14. Matsuse K, Kubota H (1991) Deadbeat flux level control of high power saturated induction servo motor using rotor flux observer. In: Conference record, IEEE-IAS, pp 409–414
15. Kobayashi H, Koizumi M, Hashimoto H, Kondo S, Harashima F (1988) A new controller for induction motors using flux observer. In: Conference record, IEEE power electronics specialists conference PESC, Atlanta, GA, pp 1063–1068
16. Verghese G, Sanders SR (1988) Observers for flux estimation in induction machines. IEEE Trans Ind Electron 35(1):85–94
17. Hori Y, Umeno T (1989) Implementation of robust flux observer based field orientation (FOFO) controller for induction machines. In: Conference record, IEEE-IAS, pp 523–528
18. Sangwongwanich S, Yonemoto T, Furuhashi T, Okuma S (1990) Design of sliding observer for robust estimation of rotor flux of induction motors. In: Proceedings of international power electronics conference (IPEC-Tokyo), pp 1235–1242
19. Kubota H, Matsuse K (1990) Flux observer of induction motor with parameter adaptation for wide speed range motor drives. In: Proceedings of international power electronics conference (IPEC-Tokyo), pp 1213–1218
20. Sangwomgwanich S, Doki S, Yonemoto T, Okuma S (1990) Adaptive sliding observers for direct field oriented control of induction motor. In: Proceedings of IEEE-IECON, pp 915–919
21. Kubota H, Matsuse k, Nakano T (1990) New adaptive flux observer of induction motor for wide speed range motor drives. In: Proceedings of IEEE-IECON, pp 921–926
22. Ben-Brahim L, Kawamura A (1991) Digital current regulation of field oriented controlled induction motor based on predictive flux observer. IEEE Trans Ind Appl 27(5):956–961
23. Bottura CP, Silvino JL, de Resende P (1993) A flux observer for induction machines based on a time-variant discrete model. IEEE Trans Ind Appl 29(2):349–354

24. Jansen PL, Lorenz RD (1994) Observer-based direct field orientation: analysis and comparison of alternative methods. IEEE IAS 30(4):945–953
25. Bose BK (1988) Technology trends in microcomputer control of electrical machines. IEEE Trans Ind Electron 35(1):160–177
26. Hori Y, Umeno T (1989) Implementation of robust flux observer based field orientation controller for induction machines. In: Conference record of the IEEE-IAS, pp 523–528
27. Kubota H, Matsuse K, Nakano T (1990) New adaptive flux observer of induction motor for wide speed range motor drives. In: Proceedings of IECON: 16th annual conference of IEEE industrial electronics society, pp 921–926
28. Bonanno GJ, Zhen L, Xu L (1995) A direct field oriented induction machine drive with robust flux estimator for position sensorless control. In: Conference record of the IEEE-IAS annual conference, pp 166–173
29. Tajima H, Hori Y (1993) Speed sensorless field orientation control of induction machine. IEEE Trans Ind Appl 2(1):175–180
30. Benlaloui I, Drid S, Chrifi-Alaoui L, Ouriagli M (2015) Implementation of a new MRAS speed sensorless vector control of induction machine. IEEE Trans Energy Convers 30(2):588–595
31. Yang G, Chin TH (1993) Adaptive speed identification scheme for a vector controlled speed sensorless inverter-induction motor drive. IEEE Trans Ind Appl 29(4):1054–1062
32. Radu B, Paolo G, Gian-Mario P (2008) Sensorless direct field-oriented control of three-phase induction motor drives for low-cost applications. IEEE Trans Ind Appl 44(2):475–481
33. Ichiroc M, Youichi O (1991) Improvement of robustness on speed sensorless vector control of induction motor. EPE Firenze 4:660–665
34. Ohtanti T, Takada N, Tanaka K (1992) Vector control of induction motor without shaft encoder. IEEE Trans Ind Appl 28(1):157–164
35. Kubota H, Matuse K, Nakano T (1993) DSP—based adaptive speed flux observer of induction motor. IEEE Trans Ind Appl 29(2):344–348
36. Kanmachi T, Takahashi I (1993) Sensorless speed control of an induction motor with no influence of secondary resistance variation. In: Conference record of the IEEE-IAS, pp 408–413
37. Kim YR, Sul SK, Park MH (1994) Speed sensorless vector control of induction motor using extended Kalman filter. IEEE Trans Ind Appl 30(5):1225–1233
38. Hurst KD, Habetler TG, Griva G, Profumo F (1994) Speed sensorless field oriented control of induction machines using current harmonic spectral estimation. In: Conference record of the IEEE IAS, vol 1, pp 601–607
39. Ilas C, Bettini A, Ferraris L, Griva G, Profumo F (1994) Comparison of different schemes without shaft sensors for field oriented control drives. In: Proceedings of IEEE-IECON, vol 3, pp 1579–1588
40. Bose BK, Simoes MG, Crecelius DR, Rajashekara K, Martin R (1995) Speed sensorless hybrid vector controlled induction motor drive. In: Conference record of the IEEE-IAS, vol 1, pp 137–143
41. Wook Kim H, Ki Sul S (1996) A new motor speed estimator using Kalman filter in low-speed range. IEEE Trans Ind Electron 43(4):498–504
42. Ropp ME, Gonzalez S (2009) Development of a MATLAB/Simulink model of a single-phase grid-connected photovoltaic system. IEEE Trans Energy Convers 24(1):195–202
43. Ministry of Electricity and Renewable Energy (New & Renewable Energy Authority (NREA)). Annual Report April 2015 of (NREA). http://www.nrea.gov.eg/annual%20report/Annual%20Report%202015.pdf
44. Metwally HM, Anis WR (1997) Performance analysis of PV pumping systems using switched reluctance motor drives. Energy Convers Manag 38(1):1–11
45. Hamrouni N, Jraidi M, Chérif A (2008) Solar radiation and ambient temperature effects on the performances of a PV pumping system. Rev Energ Renouvelables 11(1):95–106
46. Bose BK (2010) Power electronics and motor drives: advances and trends. Elsevier
47. Casadei D, Profumo F, Serra G, Tani A (2002) FOC and DTC: two viable schemes for induction motors torque control. IEEE Trans Power Electron 17(5):779–787

48. Diab AAZ (2017) Implementation of a novel full-order observer for speed sensorless vector control of induction motor drives. Elect Eng 99(3):907–921. https://doi.org/10.1007/s00202-016-0453-7
49. Rebei N, Hmidet A, Gammoudi R, Hasnaoui O (2015) Implementation of photovoltaic water pumping system with MPPT controls. Front Energy 9(2):187–198
50. Subudhi B, Pradhan R (2013) A comparative study on maximum power point tracking techniques for photovoltaic power systems. IEEE Trans Sustain Energy 4(1):89–98
51. Diab AAZ (2014) Real-time implementation of full-order observer for speed sensorless vector control of induction motor drive. J Control Autom Electr Syst 25(6):639–648. https://doi.org/10.1007/s40313-014-0149-z
52. Diab AAZ, Kotin DA, Anosov VN, Pankratov VV (2014) A comparative study of speed control based on MPC and PI-controller for indirect field oriented control of induction motor drive. In: 12th international conference on actual problems of electronics instrument engineering (APEIE), Novosibirsk, pp 728–732
53. Jin C, Li G, Qi X, Ye Q, Zhang Q, Wang Q (2016) Sensorless vector control system of induction motor by nonlinear full-order observer. In: 2016 IEEE 11th conference on industrial electronics and applications, Hefei, China, 5–7 June 2016
54. Diab AAZ, Vdovin VV, Kotin DA, Anosov VN, Pankratov VV (2014) Cascade model predictive vector control of induction motor drive. In: 12th international conference on actual problems of electronics instrument engineering (APEIE), Novosibirsk, pp 669–674
55. Munteau I, Cutululis NA, Bratcu AI, Ceanga E (2003) Optimization of variable speed wind power systems based on a LQG approach. In: Proceedings of the IFAC workshop on control applications of optimisation—CAO'03 Visegrad, Hungary, 30 June–July 2 2003
56. Hassan AA, Mohamed YS, Yousef AM, Kassem AM (2006) Robust control of a wind driven induction generator connected to the utility grid. Bull Fac Eng Assiut Univ 34:107–121
57. Attaiaence C, Perfetto A, Tomasso G (1999) Robust position control of DC drives by means of H_∞ controllers. Proc IEE Electr Power Appl 146(4):391–396
58. Naim R, Weiss G, Ben-Yakakov S (1997) H_∞ control applied to boost power converters. IEEE Trans Power Electron 12:677–683
59. Lin FJ, Lee T, Lin C (2003) Robust H_∞ controller design with recurrent neural network for linear synchronous motor drive. IEEE Trans Ind Electron 50(3):456–470
60. Pohl L, Vesely I (2016) Speed control of induction motor using H_∞ linear parameter varying controller. IFAC 49(25):74–79
61. Kao YT, Liu CH (1992) Analysis and design of microprocessor-based vector-controlled induction motor drives. IEEE Trans Ind Electron 39(1):46–54
62. Prempain E, Postlethwaite I, Benchaib A (2002) A linear parameter variant H_∞ control design for an induction motor. Control Eng Pract 10(6):633–644
63. Rigatos G, Siano P, Wira P, Profumo F (2015) Nonlinear H-infinity feedback control for asynchronous motors of electric trains. Intell Ind Syst 1(2):85–98

Chapter 3
Development and Stabilization of Adaptive State Observers for Induction Machines

Abstract In this chapter adaptive state observers that takes parameters variation into consideration is used to achieve high performance without spoiling the dynamic response. These adaptive state observers are designed using the induction machine model. A modified gain rotor flux observer with a parameter adaptive scheme has been proposed for sensorless vector control induction motor drives. The value of the observer optimal gain has been derived by minimizing the error between the measured and estimated states. Also, the stability of the proposed rotor flux observer with the parameter identification scheme is proved using the Lyapunov's theorem.

Keywords Adaptive state observers · Induction machines · Speed observers

3.1 Introduction

Recently, the sensorless vector control method of induction motor drives are widely employed in high performance industrial applications. However, this method requires exact information of the machine parameters, rotor flux and motor speed. The indirect field-oriented control method is widely used for induction motor drives. This method has a disadvantage which is sensitivity of motor parameters variations. Especially, resistance varies widely with the motor temperature. On the other hand, the direct field-oriented control method is robust against the motor parameters variations because the measured motor flux is fed back to the reference. The disadvantage of this method is that the installation of a flux sensor is necessary. Therefore, flux estimation from the terminal variables is usually used instead of the flux measurement. However, flux estimation is also sensitive to the motor parameter variations [1, 2]. To solve these problems, a proposed rotor flux observer with stator and rotor resistance adaptive scheme is presented.

In this chapter adaptive state observers that takes parameters variation into consideration is used to achieve high performance without spoiling the dynamic response. The proposed observer-based estimators have the following advantages:

© The Author(s), under exclusive license to Springer Nature Singapore Pte Ltd. 2020
A. A. Z. Diab et al., *Development of Adaptive Speed Observers for Induction Machine System Stabilization*, SpringerBriefs in Electrical and Computer Engineering,
https://doi.org/10.1007/978-981-15-2298-7_3

1. It does not use pure integration.
2. It can cope with the system nonlinearities and variations.
3. It can identify the full parameters and motor speed simultaneously.

These adaptive state observers are designed using the induction machine model. A modified gain rotor flux observer with a parameter adaptive scheme has been proposed for sensorless vector control induction motor drives. The value of the observer optimal gain has been derived by minimizing the error between the measured and estimated states. Also, the stability of the proposed rotor flux observer with the parameter identification scheme is proved using the Lyapunov's theorem.

The model reference adaptive system (MRAS) is constructed to estimate the motor speed based on the measurement of stator voltages and currents. This technique depends on the comparison between the rotor flux outputs for minimizing the resulting flux error. In this work, the stability of the estimator scheme is proved based on Popov's criterion. The estimated speed of the motor is used as a control signal in a sensor-free field-oriented control mechanism for induction motor drives.

3.2 Dynamic Model of the Induction Machine

The dynamic model of the Induction Motor is rewritten from the three-phase model of the induction motor in α-β stationary frame as [3–5]:

$$p\begin{bmatrix} i_s \\ \psi_r \end{bmatrix} = \begin{bmatrix} A_{11} & A_{12} \\ A_{21} & A_{22} \end{bmatrix}\begin{bmatrix} i_s \\ \psi_r \end{bmatrix} + \begin{bmatrix} B \\ 0 \end{bmatrix} v_s \tag{3.1}$$

Equation (3.1) can be expressed in the form of state space as:

$$px = Ax + Bv_s \tag{3.2}$$

The output equation of state space model is given by

$$i_s = Cx \tag{3.3}$$

where,

$i_s = \begin{bmatrix} i_{s\alpha} & i_{s\beta} \end{bmatrix}^T$: Stator Current,

$\psi_r = \begin{bmatrix} \psi_{r\alpha} & \psi_{r\beta} \end{bmatrix}^T$: Rotor Flux,

$V_s = \begin{bmatrix} V_{s\alpha} & V_{s\beta} \end{bmatrix}^T$: Stator Voltage,

$$A_{11} = -\left\{ \left(\frac{R_s}{\sigma L_s L_r \tau_r} - \frac{1-\sigma}{\sigma \tau_r} \right) \right\} I = a_{r11} I,$$

$$A_{12} = \left\{ \frac{L_m}{\sigma L_s L_r \tau_r} \right\} I - \left\{ \frac{L_m}{\sigma L_s L_r \tau_r} \omega_r \right\} J = a_{r12} I + a_{i12} J,$$

$$A_{21} = \left(\frac{L_m}{\tau_r} \right) I = a_{21} I,$$

$$A_{22} = -\left(\frac{1}{\tau_r} \right) I + \omega_r J = a_{r22} I + a_{i22} J,$$

$$B = \left(\frac{1}{\sigma L_s} \right) I = b_1 I,$$

$$C = [I\ 0],$$

I and J are unit matrix and skew symmetric matrix respectively:

$$I = \begin{bmatrix} 1 & 0 \\ 0 & 1 \end{bmatrix} \quad \text{and} \quad J = \begin{bmatrix} 0 & -1 \\ -1 & 0 \end{bmatrix}$$

The electromagnetic torque can be expressed in terms of stator current component and rotor current components as:

$$T_e = K_t (\psi_{r\alpha} i_{s\beta} - \psi_{r\beta} i_{s\alpha}) \tag{3.4}$$

where,

$$K_t = \frac{3 P L_m}{2 L_r}.$$

3.3 Full Order Adaptive State Observer

A full order adaptive rotor flux observer which estimates the stator current and rotor flux simultaneously can be constructed as [6]

$$p \begin{bmatrix} i_s \\ \psi_r \end{bmatrix} = \begin{bmatrix} A_{11} & A_{12} \\ A_{21} & A_{22} \end{bmatrix} \begin{bmatrix} i_s \\ \psi_r \end{bmatrix} + \begin{bmatrix} B \\ 0 \end{bmatrix} v_s + G(\hat{i}_s - i_s) \tag{3.5}$$

The stability of the proposed flux observer with the parameter adaptation scheme is proved by the Lyapunov's theorem [6–9]. The observer gains matrix (G) is given so that the observer poles are proportional to those of the induction motor. It is found that the elements of observer gain matrix (G) are similar to those given previously in the stationary frame as [9]:

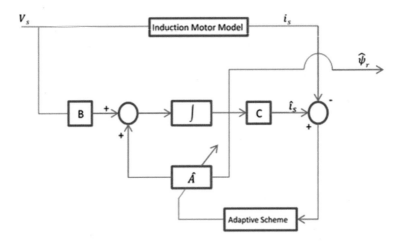

Fig. 3.1 Block diagram of the adaptive state observer for induction machine

$$G = \left[\begin{array}{cccc} g_1 & g_2 & g_3 & g_4 \\ -g_2 & g_1 & -g_4 & g_3 \end{array} \right]^T \tag{3.6}$$

where

$$g_1 = (K - 1)(\hat{a}_{r11} + \hat{a}_{22}), \quad g_2 = (K - 1)\hat{a}_{i22},$$
$$g_3 = (K^2 - 1)(c\hat{a}_{r11} + \hat{a}_{r21}) - cg_1,$$
$$g_4 = -cg_2 \quad \text{and} \quad c = \sigma \frac{L_s L_r}{L_m}$$

The full order state observer which estimates the stator current and rotor flux at the same time (Eq. (3.5)) can be rewritten in the state space form as:

$$px = Ax + BV_s + G(i_s - \hat{i}_s) \tag{3.7}$$

The block diagram of the full order adaptive state observer scheme represented by Eq. (3.7) is shown in Fig. 3.1.

3.4 Adaptive Observer Scheme for Motor Parameters and Speed Estimation

Since rotor speed ω_r is included in matrix A_{21} and A_{22}, in the second row of Eq. (3.5) can be nominated as a reference model. With replacing the actual rotor speed ω_r by the estimated value $\hat{\omega}_r$ and motor parameters keep constant. The stability of the full order observer is derived using Lyapunov theorem. The error of the estimated stator current and rotor flux based on Eq. (3.2) can be described as:

$$p \cdot e = (A + GC) \cdot e - \Delta A \hat{x} \qquad (3.8)$$

$$e = x - \hat{x},$$

$$\Delta A = A - \hat{A} = \begin{bmatrix} 0 & -\frac{\Delta \omega_r J}{C} \\ 0 & \Delta \omega_r J \end{bmatrix},$$

$$c = (\sigma L_s I_r)/L_m \quad \text{and} \quad \Delta \omega_r = \hat{\omega}_r - \omega_r$$

The following Lyapunov function applicant can be defined as:

$$V = e^T \cdot e + (\hat{\omega}_r - \omega_r)^2 / \xi \qquad (3.9)$$

where ξ is an arbitrary positive constant. The time derivative of V becomes

$$p \cdot V = e^T \{(A + GC)^2 + (A + GC)\} \cdot e - \frac{2 \Delta \omega_r (e_{is\alpha} \psi_{r\beta} - e_{is\beta} \psi_{r\alpha})}{C}$$
$$+ 2 \Delta \omega_r p \hat{\omega}_r / \xi \qquad (3.10)$$

The actual stator currents are compared to the estimated one for generating the error signal. This error is used to estimate the motor speed, stator and rotor resistances through adaptation mechanism (PI controllers) as shown in Fig. 3.2.

Where $ei_{as} = \hat{i}_{as} - i_{as}$ and $ei_{\beta s} = \hat{i}_{\beta s} - i_{\beta s}$.

The following conditions prove the stability of the adaptive rotor flux observer using Lyapunov's stability theory [10].

Observer eigenvalues have been tracked to be negative real parts, so that the states of the observer will come together with the preferred stated of the observer.

The adapted mechanism is created from the asymptotic convergences situation of the errors of the estimated speed

$$p \hat{\omega}_r = \xi (e i_{\alpha s} \hat{\psi}_{\alpha r} - e i_{\beta s} \hat{\psi}_{\beta r})/C \qquad (3.11)$$

The motor speed can be estimated according to Lyapunov's theory using PI controller as:

$$e_\omega = \left(e i_{\alpha s} \hat{\psi}_{\alpha r} - e i_{\beta s} \hat{\psi}_{\beta r} \right)$$
$$\hat{\omega}_r = e_\omega \left(K_{P\omega} + \int K_{I\omega} dt \right) \qquad (3.12)$$

Also, the stator circuit resistance which changes with motor temperature is estimated as:

$$e_{rs} = \left(e i_{\alpha s} \hat{i}_{\alpha s} + e i_{\beta s} \hat{i}_{\beta s} \right)$$

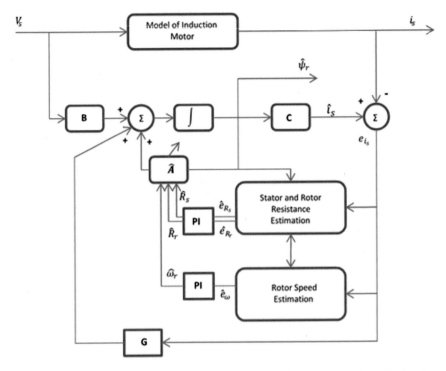

Fig. 3.2 Block diagram of full order adaptive observer scheme for parameter and speed estimations

$$\hat{R}_s = e_{rs}\left(K_{Prs} + \int K_{Irs}dt\right) \tag{3.13}$$

Moreover, the rotor circuit resistance which changes with motor temperature and frequency is estimated as:

$$e_{rr} = ei_{\alpha s}\left(\hat{\psi}_{\alpha r}\hat{i}_{\alpha s} - L_m\hat{i}_{\alpha s}\right) + ei_{\beta s}\left(\hat{\psi}_{\beta r}\hat{i}_{\beta s} - L_m\hat{i}_{\beta s}\right)$$
$$\hat{R}_r = e_{rr}\left(K_{Prr} + \int K_{Irr}dt\right) \tag{3.14}$$

where $K_{P\omega}$, $K_{I\omega}$, K_{Prs}, K_{Irs}, K_{Prr} and K_{Irr} are the PI gains of motor speed, stator and rotor resistances adaptive estimators respectively (Fig. 3.2).

3.5 Developed Gain of Adaptive Rotor Flux Observer

A rotor flux observer with a parameter adaptive scheme has been used to estimate the stator current and rotor flux for high performance applications of induction motors. However, the observer gain is chosen arbitrarily in most of the previous researches. This may have the disadvantages of large overshoot, large steady state error and increasing rise time. In this book, a modified gain rotor flux observer has been proposed for sensorless vector-controlled induction motor drives.

It is clear from Eqs. (3.5) and (3.6) that, the poles positions of the observer are assigned as K times that of the motor poles. In most literature, the observer gain K is chosen as a constant value, often equal to unity [6–9]. Therefore, the effects of the observer gain on the error dynamics and on the sensitivity to parameters variation have been neglected. To overcome these problems, the observer gain has been deduced in terms of the estimated parameters. This value is obtained based on minimizing the stator current error. This would enable to make the observer more effective.

Based on Eq. (3.5), a full order observer for estimating the stator current and rotor flux can be constructed by the following equations:

$$
\frac{d\hat{i}_{\beta s}}{dt} = \frac{L_m R_r}{\sigma L_s L_r^2} \hat{\psi}_{\alpha r} + \frac{L_m}{\sigma L_s L_r} \omega_r \hat{\psi}_{\alpha r} - \left(\frac{L_m^2 R_r + L_r^2 R_s}{\sigma L_s L_{rr}^2} \right) \hat{i}_{\alpha s} + \omega_e \hat{i}_{\beta s}
$$
$$
+ \frac{1}{\sigma L_s} V_{\alpha s} + g_1 e_{\alpha s} - g_2 e_{\beta s} \tag{3.15}
$$

$$
\frac{d\hat{i}_{\beta s}}{dt} = -\frac{L_m}{\sigma L_s L_r} \omega_r \hat{\psi}_{\alpha r} + \frac{L_m R_r}{\sigma L_s L_r^2} \hat{\psi}_{\beta r} - \omega_e i_{\alpha s} - \left(\frac{L_m^2 R_r + L_r^2 R_s}{L_s L_r^2} \right) \hat{i}_{\beta s}
$$
$$
+ \frac{1}{\sigma L_s} V_{\beta s} + g_2 e_{\alpha s} + g_1 e_{\beta s} \tag{3.16}
$$

$$
\frac{d\hat{\psi}_{\alpha r}}{dt} = -\frac{1}{\tau_r} \hat{\psi}_{\beta r} - \omega_{sl} \hat{\psi}_{\beta r} + \frac{1}{\tau_r} L_m \hat{i}_{\alpha s} + g_3 e_{\alpha s} - g_4 e_{\beta s} \tag{3.17}
$$

$$
\frac{d\hat{\psi}_{\alpha r}}{dt} = \omega_{sl} \hat{\psi}_{\beta r} - \frac{1}{\tau_r} \hat{\psi}_{\beta r} + \frac{1}{\tau_r} L_m \hat{i}_{\alpha s} + g_4 e_{\alpha s} + g_3 e_{\beta s} \tag{3.18}
$$

where $e_{\alpha s} = \hat{i}_{\alpha s} - i_{\alpha s}$ and $e_{\beta s} = \hat{i}_{\beta s} - i_{\beta s}$.

The stator voltage equations are given by:

$$
V_{\alpha s} = \frac{L_m}{L_r} \dot{\psi}_{\alpha r} + R_s i_{\alpha s} + \sigma L_s i_{\alpha s} \tag{3.19}
$$

$$
V_{\beta s} = \frac{L_m}{L_r} \dot{\psi}_{\beta r} + R_s i_{\beta s} + \sigma L_s i_{\beta s} \tag{3.20}
$$

In the field oriented control of an induction motor, the ideal decoupling between α and β axes can be achieved by letting rotor flux linkage be aligned in the α-axis, i.e.

$$\psi_{\beta r} = 0.0 \quad \text{and} \quad \dot{\psi}_{\beta r} = 0.0 \tag{3.21}$$

At steady state, all the state variable in the α-β stationary reference frame have constant dc values and their derivatives will become zero, i.e.

$$\hat{i}_{\alpha s} = \dot{\hat{i}}_{\alpha s} = 0.0 \quad \text{and} \quad \dot{\hat{i}}_{\alpha s} = \dot{\hat{i}}_{\beta s} \tag{3.22}$$

$$\dot{\psi}_{\beta r} = \dot{\hat{\psi}}_{\beta r} = \dot{\hat{\psi}}_{\alpha r} = 0.0 \tag{3.23}$$

Substituting Eqs. (3.19)–(3.23) into Eqs. (3.15)–(3.18), yields

$$\left(g_1 - \frac{R_s}{\sigma L_s}\right)e_{\alpha s} - \frac{1-\sigma}{\sigma \tau_r}\hat{i}_{\alpha s} + \omega_e \hat{i}_{\beta s} + \frac{L_m}{\sigma L_s L_r \tau_r}\hat{\psi}_{\alpha r} - g_2 e_{\beta s} = 0 \tag{3.24}$$

$$\left(g_1 - \frac{R_s}{\sigma L_s}\right)e_{\beta s} - \frac{1-\sigma}{\sigma \tau_r}\hat{i}_{\beta s} + \omega_e \hat{i}_{\alpha s} + \frac{\omega_r L_m}{\sigma L_s L_r}\hat{\psi}_{\alpha r} + g_2 e_{\alpha s} = 0 \tag{3.25}$$

$$\frac{L_m}{\tau_r}\hat{i}_{\alpha s} - \frac{1}{\tau_r}\hat{\psi}_{\alpha r} + g_3 e_{\alpha s} - g_4 e_{\beta s} = 0 \tag{3.26}$$

$$\frac{L_m}{\tau_r}\hat{i}_{\beta s} - \omega_{sl}\hat{\psi}_{\alpha r} + g_4 e_{\beta s} + g_3 e_{\beta s} = 0 \tag{3.27}$$

Assuming that the rotor flux and stator current converge to their actual values at steady state, then the field-orientation condition can be applied.

$$\hat{\psi}_{\alpha r} = L_m \hat{i}_{\alpha s} \tag{3.28}$$

Combining Eqs. (3.26) and (3.28) yield:

$$e_{\beta s} = \frac{g_3}{g_4}e_{\alpha s} \tag{3.29}$$

Substituting Eqs. (3.28) and (3.29) into (3.27). The observer stator current in the q-axis can be obtained as:

$$\hat{i}_{\beta s} = \omega_{sl}\tau_r \hat{i}_{\alpha s} - \frac{\tau_r}{L_m}\left(g_4 + \frac{g_3^2}{g_4}\right)e_{\alpha s} \tag{3.30}$$

Using Eqs. (3.29) and (3.30) to eliminate $\hat{i}_{\beta s}$ and $e_{\beta s}$ from Eq. (3.24).

$$\left(g_1 - \frac{R_s}{\sigma L_s}\right)e_{\alpha s} + \omega_e\left(\omega_{sl}\tau_r \hat{i}_{\alpha s} - \frac{\tau_r}{L_m}\left(g_4 + \frac{g_3^2}{g_4}\right)e_{\alpha s}\right) - \frac{g_2 g_3}{g_4}e_{\alpha s} = 0 \quad (3.31)$$

Thus, the current error function can be deduced as:

$$\frac{\hat{i}_{\alpha s} - i_{\alpha s}}{\hat{i}_{\alpha s}} = \frac{-\omega_e \omega_{sl}\tau_r}{\left(g_1 - \frac{R_s}{\sigma L_s}\right) - \frac{\omega_e \tau_r}{L_m}\left(g_4 + \frac{g_3^2}{g_4}\right) - \frac{g_2 g_3}{g_4}} \quad (3.32)$$

Equation (3.33) can be rewritten as:

$$\frac{\hat{i}_{\alpha s} - i_{\alpha s}}{\hat{i}_{\alpha s}} = \frac{-\omega_e \omega_{sl}\tau_r g_4}{\left(g_1 - \frac{R_s}{\sigma L_s}\right)g_4 - \frac{\omega_e \tau_r}{L_m}\left(g_4^2 + g_3^2\right) - g_2 g_3} \quad (3.33)$$

The value of the observer gain K, which minimizes the stator current error, can be obtained by differentiating the error function Eq. (3.33) with respect to K and equating to zero. Thus,

$$\frac{d}{dK}\left(\frac{\hat{i}_{\alpha s} - i_{\alpha s}}{\hat{i}_{\alpha s}}\right) = \frac{d}{dK}\left(\frac{-\omega_e \omega_{sl}\tau_r g_4}{\left(g_1 - \frac{R_s}{\sigma L_s}\right)g_4 - \frac{\omega_e \tau_r}{L_m}\left(g_4^2 + g_3^2\right) - g_2 g_3}\right) \quad (3.34)$$

Substituting the values of g_1, g_2, g_3 and g_4 into (3.34), the following quadratic equation can be obtained.

$$AK^2 + BK + C = 0 \quad (3.35)$$

where,

$$A = \frac{3\omega_e \tau_r}{L_m}(ca_{r11} + a_{r21})$$

$$B = \frac{\omega_e \tau_r}{L_m}\left(2(ca_{r11} + a_{r21})^2 - 4c(ca_{r11} + a_{r21})(ca_{r11} + a_{r22})\right)$$
$$+ 2(ca_{r11} + a_{r21})a_{i22}$$

$$C = \frac{\omega_e \tau_r}{L_m}\left(c^2 a_{i22}^2 - (ca_{r11} + a_{r22})^2 + C^2(a_{r11} + a_{r22})^2\right)$$

The optimal value of K, which minimizes the stator current error can be obtained by solving Eq. (3.35) and taking the absolute value

$$K = \left|\frac{-B + \sqrt{B^2 - 4AC}}{2A}\right| \quad (3.36)$$

3.6   Model Reference Adaptive System for Speed Estimation

3.6.1   Conventional Model Reference Adaptive System

A speed estimation scheme based on conventional model reference adaptive system (MRAS) has been proposed [11] and evaluated for speed sensorless direct field-oriented control of induction motor drives. This scheme is based on the measurement of stator voltages and currents. Figure 3.3 illustrates the basic configuration of conventional MRAS of speed estimation, which consists of a reference model (stator equation), an adjustable model (rotor equation) and an adaptive mechanism.

The motor voltages and currents are measured in a stationary reference frame (α-β axes), it is convenient to express these equations also, in the stationary frame as:

Reference model (stator equation)

$$p\begin{bmatrix} \psi_{\alpha r} \\ \psi_{\beta r} \end{bmatrix} = \frac{L_r}{L_m}\begin{bmatrix} V_{\alpha s} \\ V_{\beta s} \end{bmatrix} - \begin{bmatrix} (R_s + \sigma L_s p) & 0 \\ 0 & (R_s + \sigma L_s p) \end{bmatrix}\begin{bmatrix} i_{\alpha s} \\ i_{\beta s} \end{bmatrix} \tag{3.37}$$

Adjustable model (rotor equation)

$$p\begin{bmatrix} \hat{\psi}_{\alpha r} \\ \hat{\psi}_{\beta r} \end{bmatrix} = \begin{bmatrix} (R_s + \sigma L_s p) & -\omega_r \\ \omega_r & (R_s + \sigma L_s p) \end{bmatrix}\begin{bmatrix} \hat{\psi}_{\alpha r} \\ \hat{\psi}_{\beta r} \end{bmatrix} + \frac{L_m}{T_r}\begin{bmatrix} i_{\alpha s} \\ i_{\beta s} \end{bmatrix} \tag{3.38}$$

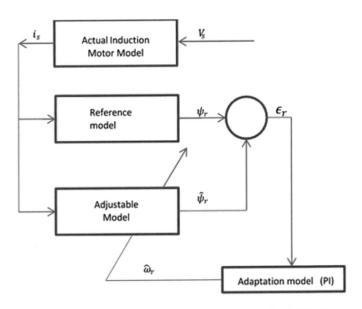

Fig. 3.3 Block diagram of motor speed estimation based on conventional MRAS

The stator equation expressed by Eq. (3.37) does not involve rotor speed ω_r, while the rotor equation expressed by Eq. (3.38) does. The conventional MRAS technique depends on the comparison between the rotor flux outputs of these two models and adjusting the value ω_r, in the rotor equation for minimizing the resulting flux error. The adjustment value of the motor speed is generated from the error between the reference and adjustable model observation by using a suitable adaptation mechanism as:

$$\hat{\omega}_r = \left(K_p + \frac{K_1}{s} \right) \varepsilon \tag{3.39}$$

where,

$$\varepsilon = \left(\hat{\psi}_{\alpha r} \psi_{\beta r} - \hat{\psi}_{\beta r} \psi_{\alpha r} \right)$$

3.6.2 Modified Model Reference Adaptive System

The α-β components of counter e.m.fs are defined as:

$$e_{m\alpha} = \frac{L_m}{L_r} p\psi_{\alpha r} \quad \text{and} \quad e_{m\beta} = \frac{L_m}{L_r} p\psi_{\beta r}$$

Thus, the counter e.m.f equations can be derived from Eq. (3.37) for stator model (reference model) as follows:

$$e_{m\alpha} = V_{\alpha s} - (R_s + \sigma L_s p)i_{\alpha s} \tag{3.40}$$

$$e_{m\beta} = V_{\beta s} - (R_s + \sigma L_s p)i_{\beta s} \tag{3.41}$$

This reference model contains the stator resistance and stator leakage inductance; therefore it is sensitive to the stator parameters variations.

Similarly, the counter e.m.f equation can be derived from Eq. (3.38) for rotor model (adjustable model) as follows:

$$\hat{e}_{m\alpha} = \frac{L_m}{L_r} \left(-\frac{1}{T_s} \psi_{\alpha r} - \omega_r \psi_{\beta r} + \frac{L_m}{L_r} i_{\alpha s} \right) \tag{3.42}$$

$$\hat{e}_{m\beta} = \frac{L_m}{L_r} \left(-\frac{1}{T_s} \psi_{\beta r} - \omega_r \psi_{\alpha r} + \frac{L_m}{L_r} i_{\beta s} \right) \tag{3.43}$$

The adjustable model is sensitive to the rotor parameters variation represented in the time constant T_r.

The measured stator voltages and currents are used as inputs for two independent estimators. These estimators can be constructed based on Eqs. (3.39) and (3.40) for the reference model and Eqs. (3.41) and (3.42) for the adjustable model in the conventional MRAS speed estimation. Note that in reference model, there is no pure integration with a subsequent robustness of the estimation at low speed. In order to eliminate the effect of stator leakage inductance, a speed estimator based on a modified MRAS has been proposed [12] and applied for speed sensorless indirect field-oriented control of induction motor drives. In this estimator, Eq. (3.39) after being multiplied by $pi_{\beta s}$ is subtracting from Eq. (3.40) after being multiplied by $pi_{\alpha s}$ to define D_e expressed by Eq. (3.43). Similarly, subtracting of Eq. (3.41) after being multiplied by $pi_{\beta s}$ from Eq. (3.42) after being multiplied by $pi_{\alpha s}$ gives $\hat{D}_e$ in Eq. (3.44).

$$D_e = \left(V_{\beta s}\,pi_{\alpha s} - V_{\alpha s}\,pi_{\beta s}\right) - R_s\left(i_{\beta s}\,pi_{\alpha s} - i_{\alpha s}\,pi_{\beta s}\right) \tag{3.44}$$

$$\hat{D}_e = \frac{L_m}{L_r}\left(\begin{array}{c} \frac{L_m}{T_r}\left(\psi_{\alpha r}\,pi_{\beta s} - \psi_{\beta r}\,pi_{\alpha s}\right) + \hat{\omega}_r\left(\psi_{\alpha r}\,pi_{\alpha s} - \psi_{\beta r}\,pi_{\beta s}\right) \\ + \frac{L_m}{T_r}\left(i_{\beta s}\,pi_{\alpha s} - i_{\alpha s}\,pi_{\beta s}\right) \end{array}\right) \tag{3.45}$$

Equation (3.43) is considered as the reference model and Eq. (3.44) as the adjustable model, of the modified MRAS speed estimator, Fig. 3.4. The leakage inductance and pure integration are eliminated in the reference model. The error between the two models of either the conventional or the modified MRAS is used

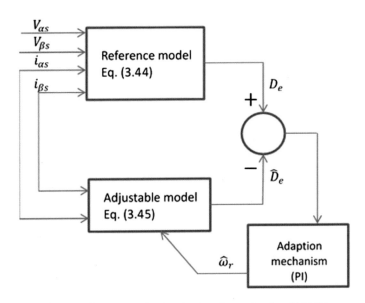

Fig. 3.4 Block diagram of motor speed estimation based on a developed MRAS

to drive a suitable adaptive mechanism, the output of which is the estimated rotor speed $(\hat{\omega}_r)$ which is used to adjust the adjustable model until the error is minimized to an acceptable limit.

References

1. Jia L, Anup T, Keith AC (2017) Speed-sensorless drive for induction machines using a novel hybrid observer. In: IEEE applied power electronics conference and exposition (APEC), pp 512–517
2. Sun K, Zhao Y (2010) A speed observer for speed sensorless control of induction motor. In: IEEE 2010 2nd international conference on future computer and communication, Wuha, China, 21–24 May 2010, vol 3, pp 263–267
3. Peter V (1990) The control of AC machines, vol 22. Oxford University
4. Lei S, Gouli L, Fang X, Bing H (2016) Speed sensorless vector control of induction motor based on adaptive full-order flux observer. In: IEEE 11th conference on industrial electronics and applications, Hefei, China, 5–7 June 2016, pp 124–127
5. Jang D-H, Choe G-H, Ehsani M (2015) Vector control of three-phase AC machines: system development in the practice. Power systems, 2nd edn. Springer
6. Kubota H, Matsuse K, Nakano T (1990) New adaptive flux observer of induction motor for wide speed range motor drives. In: IECON '90: 16th annual conference of IEEE Industrial Electronics Society, Pacific Grove, CA, USA, 27–30 Nov 1990, pp 921–926
7. Kubota H, Matsuse K, Nakano T (1993) DSP-based speed adaptive flux observer of induction motor. IEEE Trans Ind Appl 29(2):344–348
8. Matsuse K, Takokoro Y (1993) Deadbeat flux level control of direct field-oriented high horse power induction servo motor using adaptive rotor flux observer. In: IEEE industry applications conference twenty-eighth IAS annual meeting, pp 513–520
9. Yang G, Chin TH (1993) Adaptive-speed identification scheme for a vector-controlled speed sensorless inverter-induction motor drive. IEEE Trans Ind Appl 29(4):820–825
10. Suwankawin S, Sangwongwanich S (2006) Design strategy of an adaptive full-order observer for speed-sensorless induction-motor Drives-tracking performance and stabilization. IEEE Trans Ind Electron 53(1):96–119
11. Tajima H, Hori Y (1993) Speed sensorless field orientation control of induction machine. IEEE Trans Ind Appl 2(l):175–180
12. Benlaloui I, Drid S, Chrifi-Alaoui L, Ouriagli M (2015) Implementation of a new MRAS speed sensorless vector control of induction machine. IEEE Trans Energy Convers 30(2):588–595

Chapter 4
Sensorless Vector Control for Photovoltaic Array Fed Induction Motor Driving Pumping System

Abstract In this chapter, the application of sensorless vector control drives to control a photovoltaic (PV) motor pumping system has been presented. The main objective of this work is to design and analyze a single stage maximum power point tracking (MPPT) and eliminate the speed encoder. Elimination of the speed encoder aims to increase the reliability of the PV motor pumping system. Moreover, the incremental conductance method is used for MPPT of the PV system. The principles of vector control have been applied to control the single stage of VSI fed the three-phase induction motor. Furthermore, a full order observer has been designed to estimate the rotor speed of the motor pumping system. MATLAB/SIMULINK simulation has been used in order to validate the proposed algorithm.

Keywords Adaptive state observers · Induction machines · Speed observers

4.1 Introduction

The conventional pumping systems are powered from fossil fuels. The costs of fossil fuels are increased and cause additional harming effects to the environment with high pollution and greenhouse gas emissions [1–6].

In this book, the PV pumping system based on PV array, dc link capacitor, a VSI, and induction motor is proposed. The vector control is used to regulate the current of PWM VSI feeding sensorless induction motor. The elimination of the motor speed encoder and estimation of the rotor speed in sensorless induction motor drives results in a high reliability. Therefore, the full order adaptive state observer with a modified gain has been applied for estimating the rotor speed.

In this chapter, the application of sensorless vector control drives to control a photovoltaic (PV) motor pumping system has been presented. The main objective of this work is to design and analyze a single stage maximum power point tracking (MPPT) and eliminate the speed encoder. Elimination of the speed encoder aims to increase the reliability of the PV motor pumping system. Moreover, the incremental conductance method is used for MPPT of the PV system. The principles of vector

A. A. Z. Diab et al., *Development of Adaptive Speed Observers for Induction Machine System Stabilization*, SpringerBriefs in Electrical and Computer Engineering, https://doi.org/10.1007/978-981-15-2298-7_4

control have been applied to control the single stage of VSI fed the three-phase induction motor. Furthermore, a full order observer has been designed to estimate the rotor speed of the motor pumping system. MATLAB/SIMULINK simulation has been used in order to validate the proposed algorithm. The control scheme with full order observer has been simulated under different operating conditions of varying natures of solar radiation and air temperature. The simulation results demonstrate that the PV motor pumping system has a high dynamic performance with accurate estimation of motor speeds under different operating conditions.

4.2   Description of the Proposed Standalone PV Motor Pumping System

The schematic configuration of the proposed PV pumping system is shown in Fig. 4.1. In this system, the PV module followed by inverter (DC/AC) and the sensorless induction motor, to which the centrifugal pump is directly connected.

The current regulated PWM VSI is controlled by the vector control method. An adaptive speed observer is used to estimate the motor speed instead of using speed encoder.

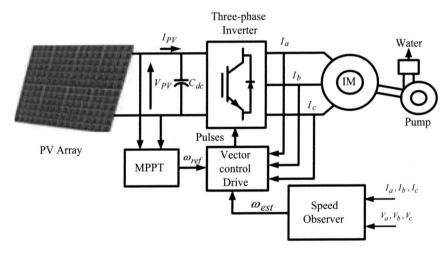

Fig. 4.1 Schematic configuration of the proposed controlled PV motor pumping system

4.3 Modeling of the Proposed Standalone PV Motor Pumping System

4.3.1 Modeling of the Photovoltaic Cell

The two-diode model representing the operation of the solar cell is proposed in these papers [7, 8]. The equivalent circuit describing the electric representation of the two-diode model has been displayed in Fig. 4.2. The output current of the cell can be expressed as the following model [7, 9–12]:

$$I = I_{PV} - I_{o1}\left[\exp\left(\frac{V - IR_s}{a_1 V_{T1}}\right) - 1\right] - I_{o2}\left[\exp\left(\frac{V - IR_s}{a_2 V_{T2}}\right) - 1\right] - \left(\frac{V - IR_s}{R_p}\right)$$

(4.1)

$$V_{T1} = V_{T2} = V_T = (N_s KT/q)$$

(4.2)

$$I_o = \frac{(I_{sc.STC} + K_I \Delta T)}{\exp[(V_{oc.STC} + K_V \Delta T)/\{(a_1 + a_2)/P\}V_T] - 1}$$

(4.3)

$$I_{o1} = I_{o2} = I_o$$

(4.4)

where, $a_2 \geq 1.2$ and $a_1 = 1$ [7, 9].

PV array is built up using series and parallel connections of PV panels for matching the required power so the previous Eq. (4.1) will become:

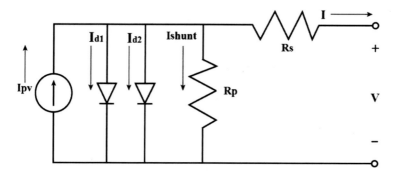

Fig. 4.2 Two-diode model equivalent circuit of solar cell

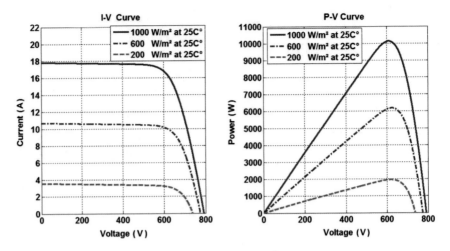

Fig. 4.3 I-V and P-V curves under variation of irradiation at STC temperature (25 °C)

$$I = N_{par} \left\{ I_{PV} - I_0 \left[\exp\left(\frac{\left(\frac{V}{N_{ser}}\right) + \left(\frac{IR_s}{N_{par}}\right)}{a_1 V_{T1}}\right) + \exp\left(\frac{\left(\frac{V}{N_{ser}}\right) + \left(\frac{IR_s}{N_{par}}\right)}{a_2 V_{T2}}\right) - 2 \right] \right\}$$
$$- \left(\frac{V\left(\frac{N_{par}}{N_{ser}}\right) - IR_s}{R_P} \right) \tag{4.5}$$

Figures 4.3 and 4.4 depict the simulation results of the photovoltaic model based on two diode model large array ($N_{ser} = 35$, $N_{par} = 2$) under variation of irradiation at STC temperature (25 °C) and variation of temperature at STC irradiation (1000 W/m^2) respectively.

4.3.2 Estimation of DC Link Capacitor

The DC bus capacitor is estimated using the expression,

$$C_{dc} \geq \frac{P}{2\omega V_{dc} \Delta v_{dc}} \tag{4.6}$$

where, V_{dc} is the reference DC voltage of VSI and P is the power consumed by the pump-motor set, ω is the line angular frequency in rad/sec and Δv_{dc} is the voltage ripple amplitude of the DC-link capacitor [13, 14]. The estimated value of the capacitor is about 3000 μF.

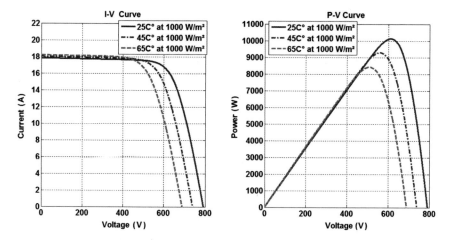

Fig. 4.4 I-V and P-V curves under variation of temperature at STC irradiation ($1000 \ \text{W/m}^2$)

4.3.3 Design of the Centrifugal Pump

From the torque-speed characteristics of the three-phase IMD directly coupled with the centrifugal pump, torque T_n is directly proportional to the square of the rotational speed ω [15].

$$T_n = K_P \times \omega_n^2 \tag{4.7}$$

where: T_n is the nominal torque of the induction motor. Also, ω_n is the nominal speed of the rotor shaft rad/sec. Consequently, the constant K_P should be:

$$K_P = T_n/\omega_n^2 \tag{4.8}$$

The output power of motor is defined as:

$$P_{out} = K_P \times \omega^3 \tag{4.9}$$

The hydraulic power requirement for the pump can be calculated by Eq. (4.10):

$$P_{hyd} = g \times Q \times H \times \rho_{water} \tag{4.10}$$

The power generated from the PV array is defined as:

$$P_{PV} = V \times I \tag{4.11}$$

The system efficiency can be calculated in Eq. (4.12)

$$\eta_{overall} = \frac{P_{pump}}{P_{PV}} * 100\% \qquad (4.12)$$

where, $g = 9.81$ m/s^2, Q = Flow rate (m^3/s), H = Net head (m), and ρ water = Density of water (kg/m^3)

4.3.4   PWM VSI Based Current Controller Scheme

In the vector control system, the induction motor is driven by a PWM transistor inverter where current feedback loops are utilized as shown in Fig. 4.5. The current controlled PWM scheme is preferred to the voltage controlled scheme because it offers substantial advantages in eliminating stator dynamics. It has the ability to provide current protection for the drive, to eliminate current unbalance. Also, the fact that the current control provides a more direct capability for controlling motor torque than the voltage control. The current controllers PWM are divided into nonlinear current controller and linear current controller schemes. The simplest form of a nonlinear current controller is the hysteresis current controller [12, 16].

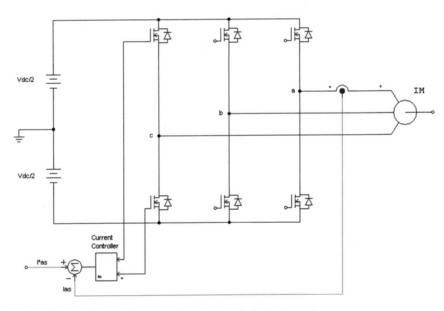

Fig. 4.5 Pulse width modulation transistor inverter for induction motor

4.3.5 Dynamic Model of Induction Motor

The dynamic model of the induction motor in α-β stationary frame is mentioned before in Eqs. (3.1)–(3.4).

4.4 Proposed Control Scheme for PV Motor Pumping System

Adaptive Incremental Conductance (AINC) is applied for MPPT. Moreover, the field-oriented control (FOC) is applied for decupling the flux and torque control loops of the induction motors. In this chapter, the MPPT algorithm is used to estimate the reference rotor speed that is referred to the output maximum power of the PV system. The reference rotor speed is the input command speed to the FOC. Moreover, the full order observer has been used to estimate the rotor speed to feed as feedback loop of the control scheme.

4.4.1 MPPT Control Algorithms

MPPT algorithms are required to extract the maximum power from a solar array [12, 17]. The incremental conductance method is depending on the fact that the slope of the power-voltage curve of the PV system is null at the point of maximum power. The slope of the curve is null at MPP, negative in the right side of MPP, and positive in left side of MPP, as shown in Fig. 4.6 [12, 17].

$$P = V \times I \ and \frac{dP}{dV} = I + V\frac{dI}{dV} \qquad (4.13)$$

At the maximum power point, Eq. 4.13 should equal zero and so;

$$\frac{dP}{dV}\Big|_{MPP} = \frac{I}{V} + \frac{dI}{dV} = 0 \qquad (4.14)$$

For any level of irradiation at the maximum power point we have:

$$\frac{dI}{dV} = \frac{-I}{V} \qquad (4.15)$$

Therefore, the reference quantity, which can be voltage, duty ratio or reference speed, is perturbed until there is a positive change in the power. The MPP can be tracked by comparing the instantaneous conductance (I/V) to the incremental conductance (ΔI/ΔV) as shown in the flowchart in Fig. 4.6. To optimize operation point

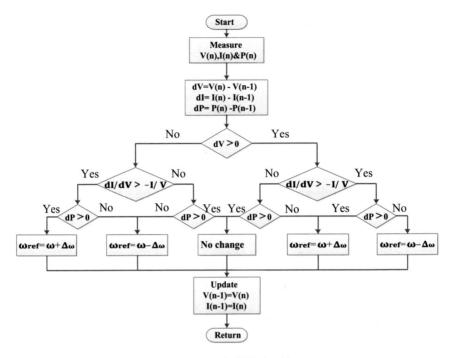

Fig. 4.6 Reference speed change according to the INC algorithm

for power extraction through controlling, the reference speed value of FOC decreased and increased to match the PV generation with the optimum motor-pump position.

4.4.2 *Field Oriented Control of Induction Motor Drive*

The vector or field-oriented control is the most commonly recognized technique of Induction motor speed control methods. When field-oriented control scheme is used, the induction motor can be competed with the DC motor in high-performance applications. This is reached by separating the three phase winding into two windings (90° apart) to simplify the control of torque and flux, which are dependent to each other. The motor speed ω_m is compared with the reference speed ω_{ref} and the error is processed by the PI speed controller which minimizes the error and produce the reference torque command T_e^* [18]. Where the actual speed can be measured by using speed sensor (tacho-generator) or can be estimated from sensorless algorithms. The reference torque T_e^* which is the output of PI controller is given by [19–22]:

$$T_e^* = K_p e + K_i \int e \, dt \qquad (4.16)$$

where, K_p, K_i are proportional and integral gains of PI controller respectively. The stator quadrature-axis current reference I_{qs}^* which corresponding to the torque can be calculated from torque reference T_e^* as:

$$I_{qs}^* = \frac{4 \times L_r}{3 \times P \times L_m \times \psi_r} \times T_e^* \tag{4.17}$$

where, L_r is the inductance of the rotor winding, P is the number of stator poles, L_m is the mutual inductance and ψ_r is the estimated rotor flux linkage given by:

$$\psi_r = \frac{L_m \cdot I_{ds}}{1 + T_r \cdot s} \tag{4.18}$$

where, $T_r = L_r/R_r$ is the rotor time constant, R_r is the rotor resistance and I_{ds} is the current command referred to the rotating d-axis. The direct-axis reference of the stator current I_{ds}^* which corresponding to the stator input flux is obtained from rotor flux reference input ψ_r^*

$$I_{ds}^* = \psi_r^* / L_m \tag{4.19}$$

At speeds higher than the rated synchronous speed of IM field weakening technique is used where I_{qs}^* increases and flux needs to be reduced in order to reduce the stator current thereby reducing the stress on the motor winding. The rotor-field angle θ_e required for coordinates transformation can be found by integrating the sum of the rotor speed ω_r and slip frequency ω_{sl}

$$\theta_e = \int (\omega_r + \omega_{sl}) dt \tag{4.20}$$

The slip frequency is calculated from the current command referred to the rotating q axis I_{qs} and the motor parameters.

$$\omega_{sl} = \frac{L_m \times R_r}{l_r \times L_r} \times I_{qs} \tag{4.21}$$

The I_{qs}^* and I_{ds}^* current references are converted into phase current references I_a^*, I_b^* and I_c^* where the reference currents I_a^*, I_b^* and I_c^* and the *measured* currents I_a, I_b and I_c are compared using hysteresis comparator to generate inverter gate signals [23]. The electromagnetic torque of the induction motor in the d-q synchronous frame can be represented as given in [24]:

$$T_e = k_t \left(\psi_{dr} i_{qs} - \psi_{qr} i_{ds} \right) \tag{4.22}$$

where, k_t is the torque constant and can be calculated as follow:

$$k_t = 3PL_m/2L_r,$$

By applying the field oriented control method, Eq. 4.22 can be described as:
$T_e = K_T i_{qs}$, where $K_T = k_t L_m i_{ds}$

In the case of constant d-axis stator current (flux current component), the electromagnetic torque is changed according to the change in the q-axis current only; and this condition can be fulfilled through closed loop control. Therefore, if the q-axis current (load current component) is projected to a rapid change according to the load current variation, the developed torque of the induction motor will be rapidly changed, and the dynamic characteristics of the motor will exhibit a high variation.

4.5 Proposed Full Order Adaptive Speed Observer

As the speed of the rotor ω_r is part of the matrix A_{12} and A_{22} of Eq. (3.1). This equation can be chosen as a reference model. When the estimated value of the rotor speed $\hat{\omega}_r$ takes place real value of rotor speed ω_r and the other parameters of the motor are kept constant, the adaptive full-order observer of speed identification system can be presented as in [19–22]:

$$p\begin{bmatrix} \hat{i}_s \\ \hat{\psi}_r \end{bmatrix} = \begin{bmatrix} A_{11} & \hat{A}_{12} \\ A_{21} & \hat{A}_{22} \end{bmatrix}\begin{bmatrix} \hat{i}_s \\ \hat{\psi}_r \end{bmatrix} + \begin{bmatrix} B \\ 0 \end{bmatrix}u_s + K\begin{bmatrix} i_s - \hat{i}_s \end{bmatrix}, \qquad (4.23)$$

where "^" refers that it is an estimated value of the variable.

According to the theory of Model Reference Adaptive System (MRAS), the induction motor described by Eq. (3.1) is considered as a reference model and the observer described by Eq. (4.23) as an adjustable model. The error feedback value is taken as the stator currents as it can be easily measured as seen in Fig. 4.7. Then, the system error e_i is taken as the difference between the measures stator current i_s and the estimated values $\hat{i}_s$.

Depending on the model of the motor given in (3.1) and the adaptive full-order observer presented in (4.22), the error equation can be expressed as [25]:

$$p\begin{bmatrix} e_i \\ e_\psi \end{bmatrix} = \begin{bmatrix} A_{11} - K \, g \, C & A_{12} \\ A'_{21} - K \, g \, C_2 & A_{22} \end{bmatrix}\begin{bmatrix} e_i \\ e_\psi \end{bmatrix} + \begin{bmatrix} 0 & \frac{-\Delta\omega_r}{\varepsilon} \, J \\ 0 & \Delta\omega_r \, J \end{bmatrix}\begin{bmatrix} \hat{i}_s \\ \hat{\psi}_r \end{bmatrix} \qquad (4.24)$$

Considering zero error in the flux, for the current error Eq. (4.24) can be re-written as following:

$$\dot{e}_i = (A_{11} - K \cdot C)e_i + \Delta A\hat{\psi}_r, \qquad (4.25)$$

where, $\varepsilon = \sigma L_s L_r / L_m$

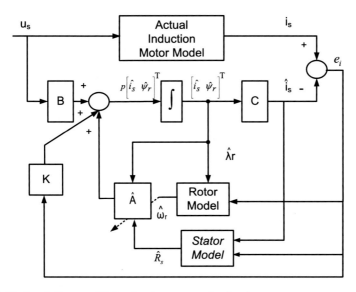

Fig. 4.7 The block diagram of full order observer speed estimation

$$\Delta A = A - \hat{A} = \begin{bmatrix} \Delta\omega_r \\ \varepsilon \end{bmatrix},$$

$$e_i = i_s - \hat{i}_s = \begin{bmatrix} e_{is\alpha} & e_{is\beta} \end{bmatrix},$$

A Lyapunov's function is selected as

$$V = e^T e + (\Delta\omega_r)^2/\psi, \tag{4.26}$$

The time derivative of V is estimated as follows:

$$\frac{dV}{dt} = \left\{ \frac{d(e^T)}{dt} \right\} e + e^T \left\{ \frac{de}{dt} \right\} + \frac{1}{\psi}\frac{d}{dt}(\Delta\omega_r)^2, \tag{4.27}$$

$$\frac{dV}{dt} = e^T\{(A-K)^T + (A-K)\}e - 2\frac{\Delta\omega_r}{\varepsilon}\left(e_{is\alpha}\hat{\psi}_{\beta r} - e_{is\beta}\hat{\psi}_{\alpha r} \right)$$
$$+ \frac{2}{\psi}\Delta\omega_r\frac{d}{dt}\hat{\omega}_r, \tag{4.28}$$

The adaptive observer stability has been accomplished using the theory of Lyapunov's stability [25], by applying the following conditions:

The observer model's eigenvalues have been chosen to have negative real parts. As a result of that, the states of the observer will converge to the proposed states of the system.

The term $\Delta\omega_r = \omega_r - \hat{\omega}_r$ included in the Eq. (4.28) has to be equal zero. A mechanism could be constructed to adapt the mechanical speed of the rotor from the asymptotic convergence's condition obtained from the estimated errors of currents.

$$\hat{\omega}_r = \frac{\lambda}{\varepsilon} \int_0^t \left\{ \hat{\psi}_{\alpha r}\left(i_{\beta s} - \hat{i}_{\beta s}\right) - \hat{\psi}_{\beta r}\left(i_{\alpha s} - \hat{i}_{\alpha s}\right) \right\} dt, \tag{4.29}$$

$$\hat{\omega}_r = \left(k_{Pes} + \frac{k_{Ies}}{s}\right) \left\{ \hat{\psi}_{\alpha r}\left(i_{\beta s} - \hat{i}_{\beta s}\right) - \psi_{\beta r}\left(i_{\alpha s} - \hat{i}_{\alpha s}\right) \right\}, \tag{4.30}$$

Therefore, according to the same Lyapunov's theory, stator resistance R_s can be estimated by a PI control equation as rotor speed by author [26, 27]

$$\hat{R}_s = \left(k_{PR} + \frac{k_{IR}}{s}\right) \left\{ \hat{i}_{\alpha s}\left(\hat{i}_{\alpha s} - i_{\alpha s}\right) - \hat{i}_{\beta r}\left(\hat{i}_{\beta s} - i_{\beta s}\right) \right\} \tag{4.31}$$

where k_{Pes}, k_{Ies}, k_{PR} and k_{IR} are PI parameters of speed and stator resistance adaptive estimators respectively, and $1/s$ is the integral operator.

4.6   Simulation Results

Matlab/Simulink model for the proposed PV pumping system under variable solar radiation (SR) and ambient temperature is presented in Fig. 4.8 where the Photovoltaic array and other blocks have been modelled mathematically. Which beneficial to calculate the reference speed at the maximum power point as the part of MPPT technique to be fed into FOC which used to drive motor and to show the characteristics curves at different temperature and irradiance levels.

The module data of the PV simulator based on two diode model at STC conditions 25 °C, AM1.5, and 1000 W/m² is listed in Appendix A. As well, the PV array design values at STC is shown in Appendix A. The induction motor coupled with the pump in the PV water pumping system under study, are capable of supplying a daily average of 47 m³/h of water at a head of 30 m [28], and its parameters is listed in Appendix A.

The performance of the PV pumping system is analyzed; in starting period, steady state, and under variation of solar irradiation level. Simulation results show that the system performs quite satisfactorily.

The characteristics of the PV pumping system under the variation of solar irradiance are shown in Fig. 4.9. The output power of PV array, the motor currents, torque, reference speed, motor speed, pump power and water flow rate pumped by the system are shown under different values of irradiation levels. This figure shows that the control system is fed power from PV array while extracting the maximum power

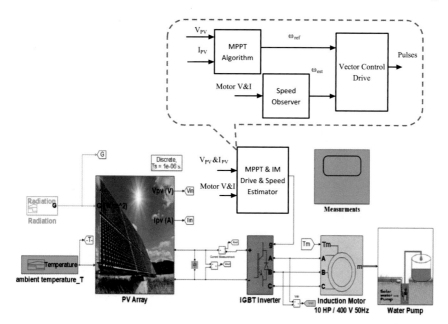

Fig. 4.8 Matlab/Simulink model of the PV pumping system under variable temperature and radiation levels

by using incremental conductance MPPT technique which calculates the reference speed to be fed into FOC which used to drive motor and determines the switching strategy for the VSI. During this period the speed will change depending on the irradiation values, since the temperature is constant at 25 °C. So when the radiation increased from zero to 900 W/m^2, the speed increased from zero to about 148 rad/s, as well as the power, voltage and current are increasing. Moreover, it can be deduced that the estimated speed and the motor speed are equal. The dynamic performance of the full order observer has been shown as well.

Figure 4.10 shows the speed error between the actual and estimated motor speed. The maximum error in the estimated speed is below 3%.

Figure 4.11 shows the characteristics of the PV pumping system with decreasing irradiation level from 900 to 600 W/m^2 at t = 1.8 s. When radiation decreases the current and the power from PV array reduces while there is not much significant drop in PV voltage. At low radiation level, maximum power that can be extracted becomes less and hence there is a reduction in the reference speed of the drive. Since, torque is a function of rotor speed, there is a dip in the torque output, thereby reducing the water flow from pump. The detailed simulation results of the pumping system performance with the full order observer have been also shown.

Fig. 4.9 The dynamic response of the PV pumping system during operation under increasing and decreasing the irradiance

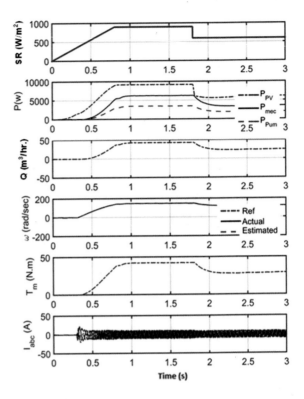

Fig. 4.10 Speed error between the actual and estimated speed

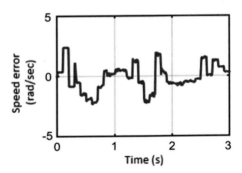

The results show the MPPT algorithm can be extract the maximum power from the PV system under variation the irradiance. The extracted power from the PV is 6000 W at irradiation of 600 W/m²; which is the maximum power at the same radiation level as per data shown in Fig. 4.3. Also, the results prove that the vector control drives have a high dynamic performance and the motor currents are balanced over the three phase. In addition, the full order observer can accurately estimate the motor speed,

Fig. 4.11 The characteristics of the system at radiation decrease from 900 to 600 W/m² at t = 1.8 s

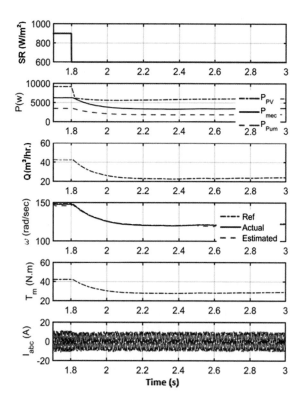

which increases the reliability of the PV pumping system with elimination of the speed encoder.

References

1. Ropp ME, Gonzalez S (2009) Development of a MATLAB/simulink model of a single-phase grid-connected photovoltaic system. IEEE Trans Energy Convers 24(1):195–202
2. Ministry of Electricity and Renewable Energy (New & Renewable Energy Authority (NREA)). Annual Report April 2015 of (NREA). http://www.nrea.gov.eg/annual%20report/Annual%20Report%202015.pdf
3. Metwally HM, Anis WR (1997) Performance analysis of PV pumping systems using switched reluctance motor drives. Energy Convers Manag 38(1):1–11
4. Hamrouni N, Jraidi M, Chérif A (2008) Solar radiation and ambient temperature effects on the performances of a PV pumping system. Rev Energ Renouvelables 11(1):95–106
5. Bose BK (2010) Power electronics and motor drives: advances and trends. Elsevier
6. Casadei D, Profumo F, Serra G, Tani A (2002) FOC and DTC: two viable schemes for induction motors torque control. IEEE Trans Power Electron 17(5):779–787
7. Ishaque K, Salam Z, Taheri H (2011) Simple, fast and accurate two-diode model for photovoltaic modules. Sol Energy Mater Sol Cells 95(2):586–594

8. Sah CT, Noyce RN, Shockley W (1957) Carrier generation and recombination in P-N junctions and P-N junction characteristics. Proc IRE 45(9):1228–1243
9. Ishaque K, Salam Z (2011) A comprehensive MATLAB simulink PV system simulator with partial shading capability based on two-diode model. Sol Energy Elsevier 85(9):2217–2227
10. Kashif I, Zainal S, Hamed T (2011) Accurate MATLAB simulink PV system simulator based on a two-diode model. J Power Electron 11(2):179–187
11. Ishaque K, Salam Z, Taheri H (2011) Modeling and simulation of photovoltaic (PV) system during partial shading based on a two-diode model. Simul Model Pract Theory Elsevier 19(7):1613–1626
12. Mahmoud MEE, Ahmed MK, Diab AAZ, Hasaneen BM (2017) Modelling and simulation of photovoltaic array fed induction motor based on field oriented control and single stage MPPT inverter for water pumping system. Journal of Multidisciplinary Engineering Science and Technology (JMEST) 4(10):8562–8569
13. Kjaer SB, Pedersen JK, Blaabjerg F (2005) A review of single-phase grid-connected inverters for photovoltaic modules. IEEE Trans Ind Appl 41(5):1292–1306
14. Zakzouk N, Abdelsalam A, Helal A, Williams B (2014) DC-link voltage sensorless control technique for single-phase two-stage photovoltaic grid-connected system. In: Energy conference (ENERGYCON), IEEE international, pp 58–64
15. Pottebaum JR (1984) Optimal characteristics of a variable-frequency centrifugal pump motor drive. IEEE Trans Ind Appl 23–31
16. Jang D-H, Choe G-H, Ehsani M (2015) Vector control of three-phase AC machines: system development in the practice. In: Power systems, 2nd edn. Springer
17. Esram T, Chapman PL (2007) Comparison of photovoltaic array maximum power point tracking techniques. IEEE Trans Energy Convers EC 22:439
18. Thangaraj R, Chelliah TR, Pant M, Abraham A, Grosan C (2011) Optimal gain tuning of PI speed controller in induction motor drives using particle swarm optimization. Log J IGPL 19(2):343–356
19. Diab AAZ (2014) Real-time implementation of full-order observer for speed sensorless vector control of induction motor drive. J Control Autom Electr Syst 25(6):639–648. https://doi.org/10.1007/s40313-014-0149-z
20. Diab AAZ, Kotin DA, Anosov VN, Pankratov VV (2014) A comparative study of speed control based on MPC and PI-controller for indirect field oriented control of induction motor drive. In: 12th international conference on actual problems of electronics instrument engineering (APEIE), Novosibirsk, pp 728–732
21. Jin C, Li G, Qi X, Ye Q, Zhang Q, Wang Q (2016) Sensorless vector control system of induction motor by nonlinear full-order observer. In: 2016 IEEE 11th conference on industrial electronics and applications, Hefei, China, 5–7 June 2016
22. Diab AAZ, Vdovin VV, Kotin DA, Anosov VN, Pankratov VV (2014) Cascade model predictive vector control of induction motor drive. In: 12th international conference on actual problems of electronics instrument engineering (APEIE), Novosibirsk, pp 669–674
23. Goyat S, Ahuja RK (2012) Speed control of induction motor using vector or field oriented control. Int J Adv Eng Technol 4(1):475
24. Diab AAZ (2017) Implementation of a novel full-order observer for speed sensorless vector control of induction motor drives. Elect Eng 99(3):907–921. https://doi.org/10.1007/s00202-016-0453-7
25. Diab AAZ, Hefnawy H, Elsayed AM (2019) Sensorless vector control drive for photovoltaic array fed induction motor with single stage MPPT inverter for water pumping system. In: Tanta university 5th international conference on scientific ISR, Sharm El Sheikh, 26–29 Mar 2019
26. Krishnan R (2001) Electric motor drives: modeling, analysis, and control. Prentice Hall
27. Mohan N (2014) Advanced electric drives: analysis, control, and modeling using MATLAB/simulink. Wiley
28. Datasheet of LORENTZ PS9k2 C-SJ42-4 (2019) http://kgelectric.co.za/assets/product_docs/lorentz/submersibles/datasheets/PSk2/LORENTZ_PS9k2_c-sj42-4_pi_en_ver301051.pdf

Chapter 5
Robust Speed Controller Design Using H∞ Theory for High Performance Sensorless Induction Motor Drives

Abstract In this chapter, a robust speed control scheme for high dynamic performance sensorless induction motor drives based on the H_infinity theory has been proposed and analyzed. Experimental and simulation results demonstrate that the presented control scheme with the H_infinity controller and MRAS speed estimator has a reasonable estimated motor speed accuracy and a good dynamic performance.

Keywords Induction machines · AC drives · H_infinity · Speed observers · Experimental and simulation validations

5.1 Introduction

The vector controlled induction motor drives are widely used in high performance industrial applications. This control system requires speed encoder which increases the cost and decreases the overall system reliability.

In this chapter, a robust speed control scheme for high dynamic performance sensorless induction motor drives based on the H_infinity (H_∞) theory has been proposed and analyzed. The proposed controller is robust against system parameter variations and achieves good dynamic performance. In addition, it rejects disturbances well and can minimize system noise. The H_∞ controller design has a standard form that emphasizes the selection of the weighting functions that achieve the robustness and performance goals of motor drives in a wide range of operating conditions. The motor speed is estimated using MRAS based on measurements of stator voltages and currents. This speed of the motor is used as a control signal in sensor-free field-oriented controlled induction motor drives. To explore the effectiveness of the suggested robust control scheme, the performance of the control scheme with the proposed controllers at different operating conditions such as a sudden change of the speed command/load torque disturbance is compared with that when using a classical controller. Experimental and simulation results demonstrate that the presented control scheme with the H_∞ controller and MRAS speed estimator has a reasonable estimated motor speed accuracy and a good dynamic performance.

© The Author(s), under exclusive license to Springer Nature Singapore Pte Ltd. 2020   49
A. A. Z. Diab et al., *Development of Adaptive Speed Observers for Induction Machine System Stabilization*, SpringerBriefs in Electrical and Computer Engineering,
https://doi.org/10.1007/978-981-15-2298-7_5

5.2   Modeling of Vector Controlled Induction Motor Drives

The induction motor can be modeled in the following mathematical differential equations represented in the rotating reference frame [1, 2]:

$$
\begin{bmatrix} V_{ds} \\ V_{qs} \\ 0 \\ 0 \end{bmatrix} = \begin{bmatrix} R_s + pL_s & -\omega_s L_s & pL_m & -\omega_s L_m \\ \omega_s L_s & R_s + pL_s & \omega_s L_m & pL_m \\ pL_m & -s\omega_s L_m & R_r pL_r & -s\omega_s L_r \\ s\omega_s L_m & pL_m & s\omega_s L_r & R_r + pL_r \end{bmatrix} \begin{bmatrix} i_{ds} \\ i_{qs} \\ i_{dr} \\ i_{qr} \end{bmatrix}
\tag{5.1}
$$

where, p is the differential operator, d/dt, L_m is the equivalent magnetizing inductance and s represents the difference between the synchronous speed and the rotor speed and it refers to slip. The self-inductances of the motor can be represented as the following:

$$
L_s = L_m + L_{ls}
$$
$$
L_r = L_m + L_{lr}
$$

where L_{ls} and L_{lr} are the stator and rotor leakage reactances, respectively.

The torque equation, in this case, is expressed as

$$
T_e = \frac{3}{2} P \frac{L_m}{L_r} \left(i_{qs} \psi_{dr} - i_{ds} \psi_{qr} \right)
\tag{5.2}
$$

The previous Eq. (5.2) is to calculate the electromagnetic torque of the motor as a function of the stator currents components, rotor flux components, pole pairs P and rotor and magnetizing inductances. Moreover, the rotor flux linkage can be written as the following equations [3]:

$$
\psi_{dr} = L_r i_{dr} + L_m i_{ds}
$$
$$
\psi_{qr} = L_r i_{qr} + L_m i_{qs}
$$

The equation of motion is

$$
J_m \frac{d\omega_r}{dt} + f_d \omega_r + T_l = T_e
\tag{5.3}
$$

where J_m is the moment of inertia, ω_r is the angular speed the rotor shaft, f_d is to express the damping coefficient, T_e indicates the electromagnetic torque of the induction motor and T_l indicates the load torque.

For achieving the finest decoupling between the ds- and qs-axis currents components, the two components of the rotor flux can be written as the following:

$$
\psi_{qr} = 0 \text{ and } \psi_{dr} = \psi_r
\tag{5.4}
$$

Based on operational requirements, when the rotor flux is set to a constant, the equation of the electromagnetic torque Eq. (5.2) will be as follows:

$$T_e = K_T i_{qs} \tag{5.5}$$

where

$$K_T = \frac{3P}{2} \frac{L_m}{L_r} \psi_r$$

Equation (5.3) can be rewritten in the s-domain as follows:

$$\omega_r = G_p(s)(T_e(s) - T_l(s)) \tag{5.6}$$

where

$$G_p(s) = \frac{1/J_m}{s + f_d/J_m} \tag{5.7}$$

A block diagram representing an indirect vector-controlled induction motor drive is shown in Fig. 5.1. The diagram consists mainly of three sub-models; a model for an induction motor under load when considering the core-loss, a hysteresis current-controlled pulse-width-modulated (PWM) inverter, and vector-control technique followed by a coordinate transformation and an outer speed-control loop.

In the vector-control scheme of Fig. 5.1, the currents i_{ds}^* and i_{qs}^* are, respectively, the magnetizing and torque current components commands.

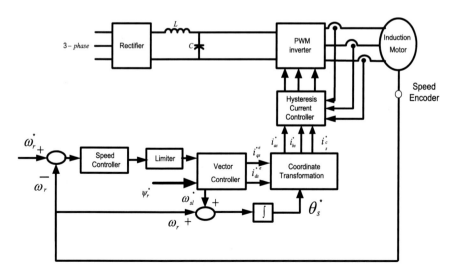

Fig. 5.1 Block diagram of an (IFO)-controlled induction motor drive

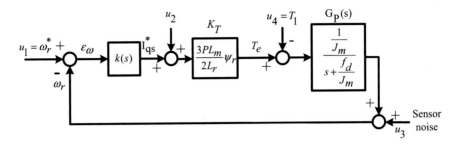

Fig. 5.2 Block diagram of the speed-control system of an IFO-controlled induction motor drive

where $I_s^* = \sqrt{i_{ds}^{*2} + i_{qs}^{*2}}$, $\theta_t^* = \tan^{-1}(i_{qs}^*/i_{ds}^*)$, and $\omega_s^* = \omega_{sl}^* + \omega_r$ (the * refers to the command value). The stator current commands of phase "a", which is the reference current command for the CRPWM inverter, is presented in references [4–7].

$$i_{as}^* = I_s^* \cos(\omega_s^* t + \theta_t^*) \tag{5.8}$$

The commands for the other two stator phases are defined below. Referring to Fig. 5.1, the slip speed command is calculated by

$$\omega_{sl}^* = \frac{1}{T_r} \frac{i_{qs}^*}{i_{ds}^*} \tag{5.9}$$

The torque current component command, i_{qs}^*, is obtained from the error of the speed, which applied to a speed controller provided that i_{ds}^* remains constant according to the operational requirements.

According to the above-mentioned analysis, the dynamic performance of the entire drive system, described in Fig. 5.1, can be represented by the control system block diagram in Fig. 5.2. This block diagram calls for accurate K_T parameters and the transfer function blocks of $G_p(s)$. In this chapter, the speed controller $K(s)$ is designed using H∞ theory to eliminate the problems inherent to classical controllers.

5.3   Design of the Proposed Robust Controller Based on H∞ Theory

The proposed controller is designed to achieve the following objectives:

(1) Minimum effect of the measurement noise at high frequency
(2) Maximum bounds on closed-loop signals to prevent saturation
(3) Minimum effect of load disturbance rejection, reducing the maximum speed dip

Fig. 5.3 General setup of
the H_∞ design problem

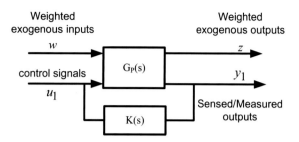

(4) Asymptotic and good tracking for sudden changes in command signals, in addition to a rapid and excellent damping response

(5) Survivability against system parameter variations.

The H_∞ theory offers a reliable procedure for synthesizing a controller that optimally verifies singular value loop-shaping specifications [8–14]. The standard setup of the H_∞ control problem consists of finding a static or dynamic feedback controller such that the H_∞ norm (a standard quantitative measure for the size of the system uncertainty) of the closed-loop transfer function below a given positive number under the constraint that the closed-loop system is internally stable.

H_∞ synthesis is performed in two stages:

i. *Formulation*: the first stage is to select the optimal weighting functions. The proper selections of the weighting functions give the ability to improve the robustness of the system at different operation condition and varying the model parameters. Moreover, this to reject the disturbance and noises besides the parameter uncertainties.

ii. *Solution*: The transfer function of the weights has been updated to reach the optimal configuration. In this chapter, the MATLAB optimization toolbox in the Simulink is used to determine the best weighting functions.

Figure 5.3 illustrates the block diagram of the H_∞ design problem, where G(s) is the transfer function of the supplemented plant [nominal plant $G_p(s)$] plus the weighting functions that represent the design features and objectives. U_1 is the control signal and w is the exogenous input vector, which generally comprises the command signals, perturbation, disturbance, noise and measurement interference; and Y_1 is the controller inputs such as commands, measured output to be controlled, and measured disturbance; its components are typically tracking errors and filtered actuator signal; and Z is the exogenous outputs; "error" signals to be minimized.

The objective of this problem is to design a controller $K(s)$ for the augmented plant G(s) to have desirable characteristics for the input/output transfer based on the information of Y_1 [inputs to the controller $K(s)$] to generate the control signal u1. Therefore, the design and selection of the $K(s)$ should counteract the influence of w and z. As a conclusion, the H_∞ design problem can be subedited as detecting an equiponderating feedback control law $u1$ (s) $= K(s) y_1(s)$ to neutralizes the effect of w and z and so to minimize the closed loop norm from w to z.

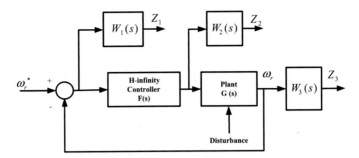

Fig. 5.4 Simplified block diagram of the augmented plant, including the H∞ controller

In the proposed control system that includes the H∞ controller, one feedback loop is designed to adjust the speed of the motor, as given in Fig. 5.4. The nominal system $G_P(s)$ is augmented with the weighting transfer functions $W_1(s)$, $W_2(s)$ and $W_3(s)$, which penalize the error signals, control signals, and output signals, respectively. The selection of the appropriate weighting functions is the quintessence of the H∞ control. The wrong weighting function may cause the system to suffer from poor dynamic performance and instability characteristics.

Consider the augmented system shown in Fig. 5.3. The following set of weighting transfer functions are selected to represent the required robustness and operation objectives:

A good choice for $W_1(s)$ is helpful for achieving good input reference tracking and good disturbance rejection. The matrix of the weighted error transfer function Z_1, which is needed for regulation, can be driven as follows:

$$Z_1 = W_1(s)\left[\omega_{ref} - \omega_r\right].$$

A proper selection for the second weight $W_2(s)$ will assist in excluding actuator saturation and provide robustness to plant supplemented disturbances. The matrix of the weighted control function Z_2 can be expressed as:

$$Z_2 = W_2(s).u(s),$$

where $u(s)$ is the transfer function matrix of the control signal output of the H∞ controller.

Additionally, a proper selection for the third weight $W_3(s)$ will restrict the bandwidth of the closed loop and achieve robustness to plant output multiplicative perturbations and sensor noise attenuation at high frequencies. The weighted output variable can be provided as:

$$Z_3 = \omega_r W_3(s).$$

In summary, the transfer functions of interest that determine the behavior of the voltage and power closed-loop systems are:

(a) Sensitivity function: $S = [I + G(s) \cdot K(s)]^{-1}$,

where G(s) and F(s) are the transfer functions of the nominal plant and the H$_\infty$ controller, respectively, while I is the identity matrix. Therefore, when S is minimized at low frequencies, it will secure perfect tracking and disturbance rejection.

(b) Control function: $C = K(s)[I + G(s) \cdot K(s)]^{-1}$.

Minimizing C will preclude saturation of the actuator and acquire robustness to plant additional disturbances.

(c) Complementary function: $T = I - S$.

Minimizing T at high frequencies will ensure robustness to plant output multiplicative perturbations and achieve noise attenuation.

5.4 Robust Speed Estimation Based on MRAS Techniques for an IFO Control

The using of speed encoder in induction machines drives spoils the ruggedness and simplicity of the induction motor. Moreover, the speed sensor increases the cost of the induction motor drives. To eliminate the speed sensor, the calculation of the speed may be based on the coupled circuit equations of the motor [15–27]. The following explanation and analysis of the stability of the model reference adaptive system (MRAS) speed estimator. In this work, the stability estimator is proven based on Popov's criterion. The measured stator voltages and currents have been used in a stationary reference frame to describe the stator and rotor models of the induction motor. The voltage model (stator model) and the current model (rotor equation) can be written as the following in the stationary reference frame $\alpha - \beta$ [15–24]:
The voltage model (stator equation):

$$p\begin{bmatrix} \psi_{\alpha r} \\ \psi_{\beta r} \end{bmatrix} = \frac{L_r}{L_m}\left(\begin{bmatrix} V_{\alpha s} \\ V_{\beta s} \end{bmatrix} - \begin{bmatrix} (R_s + \sigma L_s p) & 0 \\ 0 & (R_s + \sigma L_s p) \end{bmatrix}\begin{bmatrix} i_{\alpha s} \\ i_{\beta s} \end{bmatrix}\right) \quad (5.10)$$

The current model (rotor equation):

$$p\begin{bmatrix} \psi_{\alpha r} \\ \psi_{\beta r} \end{bmatrix} = \begin{bmatrix} (-1/T_r) & -\omega_r \\ \omega_r & (-1/T_r) \end{bmatrix}\begin{bmatrix} \psi_{\alpha r} \\ \psi_{\beta r} \end{bmatrix} + \frac{L_m}{T_r}\begin{bmatrix} i_{\alpha s} \\ i_{\beta s} \end{bmatrix} \quad (5.11)$$

Figure 5.5 illustrates an alternative way of observing the rotor speed using MRAS. Two independent rotor flux observers are constructed to estimate the components of the rotor flux vector: one based on Eq. (5.10) and the other based on Eq. (5.11). Because Eq. (5.10) does not involve the quantity ω_r, this observer may be regarded

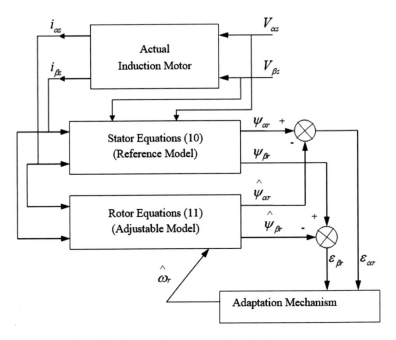

Fig. 5.5 Structure of the MRAS system for motor speed estimation

as a reference model of the induction motor, while Eq. (5.11), which does involve ω_r, may be regarded as an adjustable model. The states of the two models are compared and the error between them is applied to a suitable adaptation mechanism that produces the observed $\hat{\omega}_r$ for the adjustable model until the estimated motor speed tracks well against the actual speed.

When eliciting an adaptation mechanism, it is adequate to initially act as a constant parameter of the reference model. By subtracting Eq. (5.11) for the adjustable model from the corresponding equations belonged to the reference model Eq. (5.10) for the rotor equations, the following equations for the state error can be obtained:

$$p\begin{bmatrix} \varepsilon_{\alpha r} \\ \varepsilon_{\beta r} \end{bmatrix} = \begin{bmatrix} (-1/T_r) & -\omega_r \\ \omega_r & (-1/T_r) \end{bmatrix}\begin{bmatrix} \varepsilon_{\alpha r} \\ \varepsilon_{\beta r} \end{bmatrix} + \begin{bmatrix} -\hat{\psi}_{\beta r} \\ \hat{\psi}_{\alpha r} \end{bmatrix}(\omega_r - \hat{\omega}_r) \qquad (5.12)$$

that is,

$$p[\varepsilon] = [A_r][\varepsilon] - [W] \qquad (5.13)$$

Because $\hat{\omega}_r$ is a function of the state error, Eqs. (5.12–5.13) represent a non-linear feedback system, as shown in Fig. 5.6.

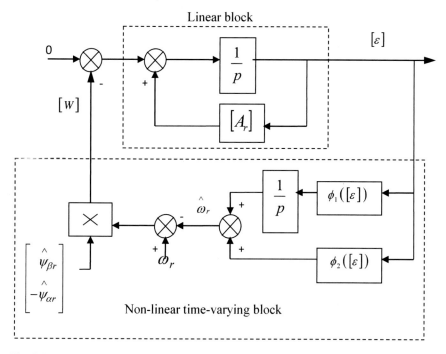

Fig. 5.6 Representation of MRAS as a non-linear feedback system

According to Landau, hyperstability is confirmed as long as the linear time-invariant forward-path transfer matrix is precisely positive real and that the non-linear feedback (which includes the adaptation mechanism) comply with Popov's hyperstability criterion. Popov's criterion demands a bounded negative limit on the input/output inner product of the non-linear feedback system. Assuring this criterion leads to the following candidate adaptation mechanism [26–29]:

Let

$$\hat{\omega}_r = \phi_2([\varepsilon]) + \int\limits_0^t \phi_1([\varepsilon])d\tau \tag{5.14}$$

then, Popov's criterion presupposes that

$$\int\limits_0^{t_1} [\varepsilon]^T [W]dt \geq -\gamma_0^2 \quad \text{for all } t_1 \geq 0 \tag{5.15}$$

where γ_0^2 is a positive constant. Substituting for $[\varepsilon]$ and $[W]$ in this inequality using the definition of $\hat{\omega}_r$, Popov's criterion for the system under study will be

$$\int_0^{t_1} \left\{ \left[\varepsilon_{\alpha r} \psi_{\beta r} - \varepsilon_{\beta r} \psi_{\alpha r} \right] \left[\omega_r - \phi_2([\varepsilon]) - \int_0^t \phi_1([\varepsilon]) d\tau \right] \right\} dt \geq -\gamma_0^2 \qquad (5.16)$$

A proper solution to this inequality can be realized via the following well-known formula:

$$\int_0^{t_1} k(pf(t)) f(t) dt \geq -\frac{1}{2} k \, f(0)^2, \quad k > 0 \qquad (5.17)$$

Using this expression, Popov's inequality is satisfied by the following functions:

$$\phi_1 = K_P \left(\varepsilon_{\beta r} \hat{\psi}_{\alpha r} - \varepsilon_{\alpha r} \hat{\psi}_{\beta r} \right) = K_P \left(\psi_{\beta r} \hat{\psi}_{\alpha r} - \psi_{\alpha r} \hat{\psi}_{\beta r} \right)$$

$$\phi_2 = K_I \left(\varepsilon_{\beta r} \hat{\psi}_{\alpha r} - \varepsilon_{\alpha r} \hat{\psi}_{\beta r} \right) = K_I \left(\psi_{\beta r} \hat{\psi}_{\alpha r} - \psi_{\alpha r} \hat{\psi}_{\beta r} \right).$$

Figure 5.7 illustrates the block diagram of the MRAS. The outputs of the two models the rotor flux components. Moreover, the measured stator voltages and currents are in the stationary reference frame have been applied to be as the inputs of MRAS.

Fig. 5.7 Block diagram of the (MRAS) speed estimation system

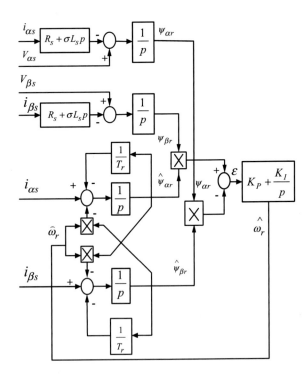

5.5 Proposed Sensor-Free Induction Motor Drive for High-Performance Applications

Figure 5.8 shows a block diagram of the proposed controller sensor-free induction motor drive system. The control system is composed of a robust controller based on H_∞ theory, hysteresis controllers, a vector rotator, a digital pulse with modulation (PWM) scheme for a transistor bridge voltage source inverter (VSI) and a motor speed estimator based on MRAS. The speed controller makes speed corrections by assessing the error between the command and estimated motor speed. The speed controller is used to generate the command q-component of the stator current i_{qs}^*. The vector rotator and phase transform in Fig. 5.8 are used to transform the stator current components command to the three-phase stator current commands $(i_{as}^*, i_{bs}^*$ and $i_{cs}^*)$ using the field angle position $\hat{\theta}_r$. The field angle is obtained by integrating the summation of the estimated speed and slip speed. The hysteresis current control compares the stator current commands to the actual currents of the machine and switches the inverter transistors in such a way as to obtain the desired command currents. Moreover, the MRAS rotor speed estimator can observe the rotor speed $\hat{\omega}_r$

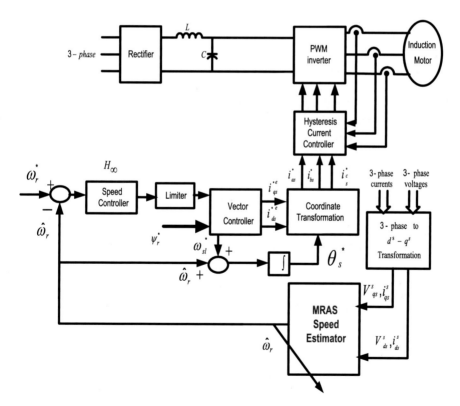

Fig. 5.8 The block diagram of the proposed sensorless induction motor drive based on H_∞ theory

based its inputs of measured stator voltages and currents. This estimated speed is fed back to the speed controller. Additionally, the estimated speed is added to the slip speed, and the sum is integrated to obtain the field angle θ_s.

5.6 Simulation and Experimental Results

5.6.1 Simulation Results and Discussions

The proposed system shown in Fig. 5.8 has been simulated using MATLAB/Simulink under different operating conditions. The parameters and data specifications of the motor used in the simulation are listed in Appendix B, Table B.1. The following set of weighting functions is obtained based on the application of the optimization toolbox in MATLAB/Simulink to achieve the proposed robustness and operation objectives:

$$W_1 = \gamma_1 \frac{s + 0.0001}{s + 0.001}, \quad W_2 = \gamma_2 \frac{s + 0.0001}{s + 10} \text{ and } W_3 = \gamma_3,$$

where, $\gamma_1 = 12$, $\gamma_2 = 0.00001$ and $\gamma_3 = 0$.

The transient behavior of the proposed sensorless control system is evaluated by applying and removing the motor-rated torque (1 N m), as shown in Fig. 5.9. Figure 5.9a shows the performance of the proposed control scheme with H∞ controller. While Fig. 5.9b illustrates the performance of sensorless induction motor (IM) drive based on the conventional PI controller for the comparison purpose.

Figure 5.9 shows that the motor speed can be effectively estimated and accurately tracks the actual speed when using the proposed sensorless scheme. Moreover, the figure shows that the performance of the two controllers of the PI controller and the proposed H∞ controller have acceptable dynamic performance. Furthermore, the figure also indicates that a fast and precise transient response to motor torque is achieved with the H∞ controller. Additionally, the stator phase current matches the value of the application and removal of the motor-rated torque.

For more validating of the H∞ controller, Fig. 5.10 describes the dynamic response of the control scheme based on the proposed H∞ controller versus the PI controller when the system is subjected to a step change in the load torque. From the figure, the results show an acceptable dynamic performance of the control scheme with the PI controllers and the H∞ against the step change in the load torque. However, the application of the H∞ controller causes improvements in the dynamic response of the rotor speed. Clearly, the PI controller requires a long rise time compared to the H∞ controller and the speed response has a larger overshoot with respect to the H∞ controller. The reasons for these results of the priority of the H∞ based induction motor drive may be because the H∞ controller has been designed taking the weighting function for the disturbance.

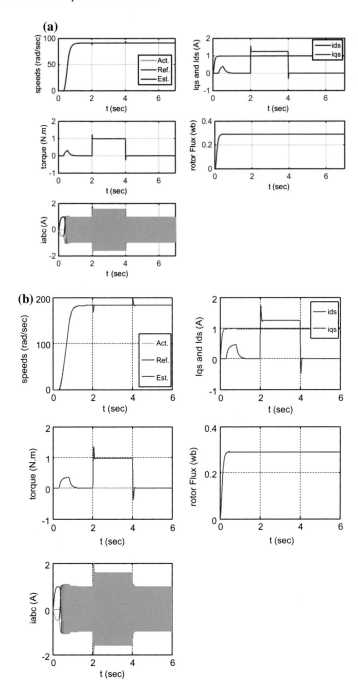

Fig. 5.9 The dynamic response of estimated motor speed, motor torque, phase currents of stator current components and rotor flux for load step changes **a** with H_∞ controller and **b** with PI controller

Fig. 5.10 Estimated motor speed response for load torque disturbance using proposed H_∞ controller compared to PI controller

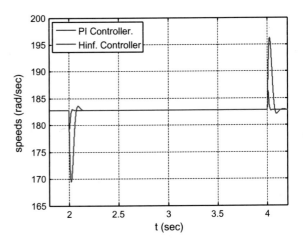

The discussion of these results can be more illustrated as follows: The fixed parameter controllers such as PI controllers are developed at nominal operating points. However, it may not be suitable under various operating conditions. However, the real problem in the robust nonlinear feedback control system is to synthesize a control law which maintains system response and error signals to within prespecified tolerances despite the effects of uncertainty on the system. Uncertainty may take many forms but among the most significant are noise, disturbance signals, and modeling errors. Another source of uncertainty is unmodeled nonlinear distortion. Consequently, researchers have adopted a standard quantitative measure the size of the uncertainty, called the H_∞. Therefore, the dynamic performance of the control scheme with the H_∞ controller is improved rather than the acceptable performance with the PI controller.

Figure 5.11 shows the actual and estimated speed transient, motor torque and stator phase current during acceleration and deceleration at different speeds. The estimated speed agreed satisfactorily with the actual speed. Additionally, a small deviation occurs from the actual speed before reaching steady-state and subsequently tracking quickly towards the command value. The motor torque response exhibits good dynamic performance. Figure 5.11 also shows the stator phase current during acceleration. From these simulation results, the proposed sensorless drive system is capable of operating at high speed, as illustrated in Fig. 5.11.

In the last case of the study, the transient performance of the sensorless induction machine drive is examined by reversing the rotor speed. Figure 5.12 shows that induction motor drive based on the proposed robust controller has high dynamic performance at reversing the rotor speed from 150 rad/sec. Moreover, the speed estimator MRAS can accurately observe the rotor speed.

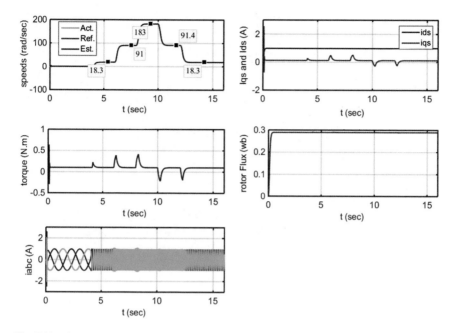

Fig. 5.11 The dynamic response of estimated motor speed, motor torque, phase currents of stator current components and rotor flux for acceleration and deceleration operation

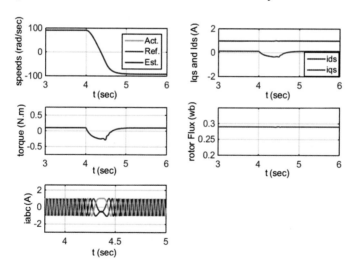

Fig. 5.12 The dynamic response of estimated motor speed, motor torque, phase currents of stator current components and rotor flux for acceleration and deceleration operation during reversing the speed

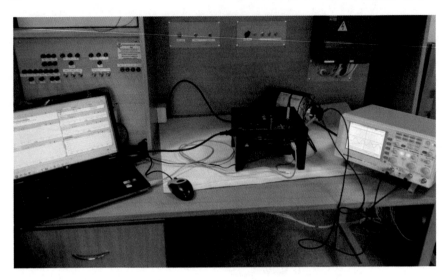

Fig. 5.13 A photo of the experimental setup for induction motor drive

5.6.2   Experimental Results and Discussions

Figure 5.13 shows the experimental setup for the configured drive system. The drive system includes an induction motor; with the same parameters and data specification of the induction motor which used for simulating the proposed control scheme; linked with a digital control board (TMDSHVMTRPFCKIT from Texas Instruments with a TMS320F28035 control card) [27, 30, 31]. The complete control scheme has been programmed in the package of Code Composer Studio CCS from Texas Instruments.

To validate the effectiveness of the sensorless vector control of the induction motor drive, the experiments were accomplished at different values of the reference speed. Figures 5.14, 5.15, 5.16 show samples of the results when the proposed system is tested at reference speeds of 0.2 pu and 0.4 pu; the base speed is assumed to be 3600 rpm (So, when the reference speed is 0.5 pu, it means the speed equals 1800 rpm). The results proved that the drive system effectively works at an extensive range of speeds. In addition, the actual and estimated speeds have coincided. Moreover, from Fig. 5.16, it is obviously seen that the current of the motor is sinusoidal.

Another case of study has been tested for more evaluating of the control scheme. In this case of study, the reference speed has been reversed from 0.4 pu to 0.2 pu in the reverse direction in a ramp variation. The results of this case of study are shown in Figs. 5.17 and 5.18. Figure 5.17 shows the speed response of the control scheme. Moreover, the phase current is shown in Fig. 5.18. The results show the control scheme has a good dynamic performance.

The last case of study has assumed many ramp changes in the references including reversing the speed. The rotor speed response is shown in Fig. 5.19. The results have

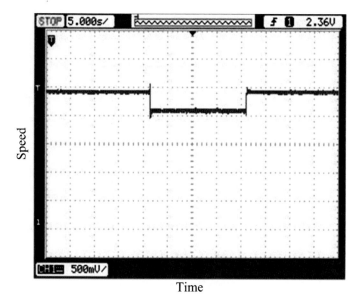

Fig. 5.14 Experimental transient response of estimated motor speed for the proposed drive system under step change of speed from 0.4 pu to 0.2 pu to 0.4 pu

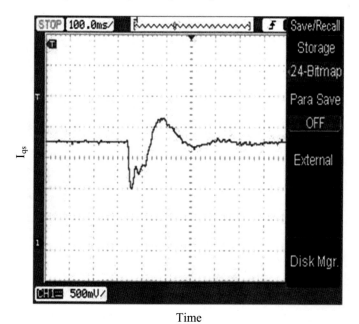

Fig. 5.15 Experimental transient response of the stator current components (i_{qs}) under step change of speed command from 0.4 pu to 0.2 pu

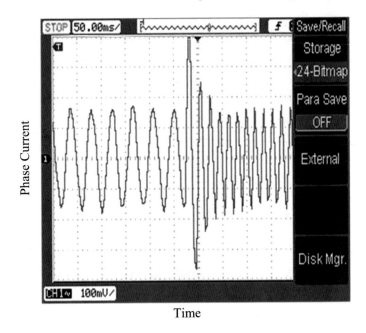

Fig. 5.16 Expremintal transient response of stator phase current (**a**) at reference speeds of 0.2 pu to 0.4 pu

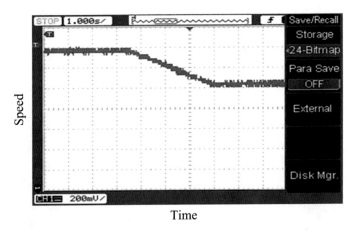

Fig. 5.17 Experimental transient response of the estimated motor speed when the speed command is changed from 0.4 pu to 0.2 pu in linear fashion

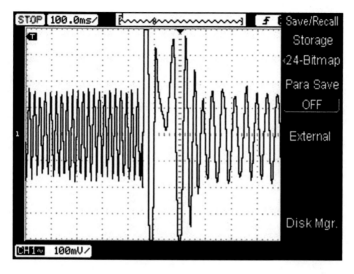

Fig. 5.18 Experimental of transient response of the stator phase current (phase a) motor speed when the speed command is changed from 0.4 pu to 0.2 pu in linear fashion

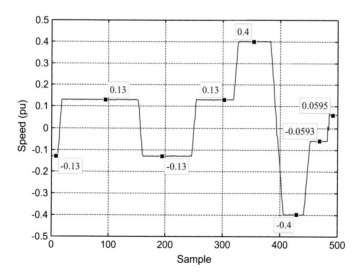

Fig. 5.19 The transient performance of multi-variation in the motor speed including speed reversal

been plotted with the aid of CCS package and digital signal processing (DSP). The experimental data and measurements have been collected with the aid of the DSP and plotted using Matlab plot tool as shown in Fig. 5.19. The results show that the control scheme has a good dynamic performance of the estimated motor speed.

References

1. Godoy M, Bose BK, Spiegel RJ (1996) Design and performance evaluation of a fuzzy-logic-based variable-speed wind generation system. IEEE Trans Energy Convers 1:349–356
2. Munteau I, Cutululis NA, Bratcu AI, Ceanga E (2003) Optimization of variable speed wind power systems based on a LQG approach. In Proceedings of the IFAC workshop on control applications of optimisation—CAO'03 Visegrad, Hungary, 30 June–2 July, 2003
3. Hassan AA, Mohamed YS, Yousef AM, Kassem AM (2006) Robust control of a wind driven induction generator connected to the utility grid. Bull Fac Eng Assiut Univ 34:107–121
4. Vas P (1993) Parameter estimation, condition monitoring, and diagnosis of electrical machines. Oxford Science Publications, Oxford, UK, vol 27
5. Toliyat HA, Levi E, Raina M (2003) A review of RFO induction motor parameter estimation techniques. IEEE Trans Energy Convers 18(2):271–283
6. Minami K, Veles-Reyes M, Elten D, Verghese GC, Filbert D (1991) Multi-stage speed and parameter estimation for induction machines. In: Proceedings of the record 22nd annual IEEE power electronics specialists conference (IEEE PESC'91), Cambridge, MA, USA, 24–27 June 1991, pp 596–604
7. Roncero-Sánchez PL, García-Cerrada A, Feliú (2007) Rotor-resistance estimation for induction machines with indirect-field orientation. Control Eng Pract 15(9):1119–1133
8. Attaiaence C, Perfetto A, Tomasso G (1999) Robust position control of DC drives by means of H_∞ controllers. Proc IEE Electr Power Appl 146(4):391–396
9. Naim R, Weiss G, Ben-Yakakov S (1997) H_∞ control applied to boost power converters. IEEE Trans Power Electron 12:677–683
10. Lin FJ, Lee T, Lin C (2003) Robust H_∞ controller design with recurrent neural network for linear synchronous motor drive. IEEE Trans Ind Electron 50(3):456–470
11. Pohl L, Vesely I (2016) Speed control of induction motor using H_∞ linear parameter varying controller. IFAC 49(25):74–79
12. Kao YT, Liu CH (1992) Analysis and design of microprocessor-based vector-controlled induction motor drives. IEEE Trans Ind Electron 39(1):46–54
13. Prempain E, Postlethwaite I, Benchaib A (2002) A linear parameter variant H_∞ control design for an induction motor. Control Eng Pract 10(6):633–644
14. Rigatos G, Siano P, Wira P, Profumo F (2015) Nonlinear H-infinity feedback control for asynchronous motors of electric trains. Intell Ind Syst 1(2):85–98
15. Riberiro LAD, Lima AMN (1999) Parameter estimation of induction machines under sinusoidal PWM excitation. IEEE Trans Energy Convers 14(4):1218–1223
16. Moonl S, Keyhani A (1994) Estimation of induction machines parameters from standstill time-domain data. IEEE Trans Ind Appl 30(6):1609–1615
17. Lima AMN, Jacobina CB, Filho EBD (1997) Nonlinear parameter estimation of steady-state induction machine models. IEEE Trans Ind Electron 44(3):390–397
18. Kubota H, Matsuse K (1994) Speed sensorless field-oriented control of induction motor with rotor resistance adaptation. IEEE Trans Ind Appl 30(5):1219–1224
19. Al-Tayie J, Acarnley P (1997) Estimation of speed, stator temperature and rotor temperature in cage induction motor drive using the extended Kalman filter algorithm. Proc IEE Electr Power Appl 144(5):301–309
20. Barut M, Bogosyan OS, Gokasan M (2007) Switching EKF technique for rotor and stator resistance estimation in speed sensorless control of induction motors. Energy Convers Manag 48(12):3120–3134
21. Jingchuan L, Longya X, Zhang Z (2005) An Adaptive Sliding-Mode Observer for Induction Motor sensorless Speed Control. IEEE Trans Ind Appl 41(4):1039–1146
22. Rashed M, Stronach AF (2004) A stable back-EMF MRAS-based sensorless low speed induction motor drive insensitive to stator resistance variation. IEE Proc Electr Power Appl 151(6):685–693

23. Mohamed YS, El-Sawy AM, Diad AAZ (2009) Rotor resistance identification for speed sensorless vector controlled induction motor drives taking saturation into account. J Eng Sci Assiut Univ 37:393–412
24. Middeton RH, Goodwin GC (1990) Digital control and estimation, 1st edn, vol 1. Prentice-Hall, Inc., Englewood Cliffs, NJ, USA,
25. Blasco-Gimenez R, Asher G, Summer M, Bradley K (1996) Dynamic performance limitations for MRAS based sensorless induction motor drives. Part 1: Stability analysis for the closed loop drive. Proc IEE-Electr Power Appl 143(2):113–122
26. Diab AAZ, Khaled A, Elwany MA, Hassaneen BM (2016) Parallel estimation of rotor resistance and speed for sensorless vector controlled induction motor drive. In: Proceedings of the 2016 17th international conference of young specialists on micro/nanotechnologies and electron devices (EDM), Erlagol, Russia, 30 June–4 July, 2016
27. Diab AAZ (2017) Novel robust simultaneous estimation of stator and rotor resistances and rotor speed to improve induction motor efficiency. Int J Power Electron 8(4):267–287
28. Diab AAZ (2017) Implementation of a novel full-order observer for speed sensorless vector control of induction motor drives. Elect Eng 99(3):907–921. https://doi.org/10.1007/s00202-016-0453-7
29. Diab AAZ (2014) Real-time implementation of full-order observer for speed sensorless vector control of induction motor drive. J Control Autom Electr Syst 25(6):639–648. https://doi.org/10.1007/s40313-014-0149-z
30. Texas Instruments C2000 Systems and Applications Team. High voltage motor control and PFC (R1.1) Kit Hardware Reference Guide, v. 2; http://www.ti.com/tool/TMDSHVMTRPFCKIT, March, 2019
31. Diab AAZ, El-Sayed A-HM, Abbas HH, Sattar MAE (2019) Robust speed controller design using H_infinity theory for high-performance sensorless induction motor drives. MDPI, ST Alban-Anlage 66, Basel, Switzerland, CH-4052, Energies, vol 12, No. 5, pp 961 (2019)

Chapter 6
Conclusions and Recommendation for Future Work

Abstract This book presents proposed adaptive rotor flux observers in order to identifying motor speed and parameter for sensorless vector control induction machine drives. The stability of this observer with the parameter adaptation scheme is derived by Lyapunov's theorem. The proposed systems have been validated through simulation and experimental testes with different applications. In view of the presented analysis and investigations, this chapter introduces the main conclusions of the work.

Keywords Induction motor drives · Field-oriented control · Speed observers · H_infinity · Experimental implementation

6.1 Conclusions

In view of the presented analysis and investigations, one can draw the following main conclusions:

1. The proposed adaptive rotor flux observers are designed based on induction machine model in order to identifying motor speed and parameter for sensorless vector control induction machine drives. The stability of this observer with the parameter adaptation scheme is derived by Lyapunov's theorem.
2. The optimal value of the modified gain rotor flux observer with a parameter adaptation scheme has been presented for sensorless vector control induction machine drives.
3. The model reference adaptive system (MRAS) estimator is constructed based on stator and rotor models to estimate the motor speed for sensorless indirect vector control induction machine drives. Also, the stability of this estimator is proved by Popov's criterion.
4. The proposed full order speed observer has been applied for photovoltaic motor pumping system driven by vector controlled VSI is presented.

5. The characteristics of photovoltaic motor pumping system with the proposed speed observer under different values of radiation are presented.
6. The estimated motor speed is approximate proportionally increased linearly as the radiation increases linearly as well as the power, voltage and current will increase.
7. The adaptive stator resistance estimation scheme is capable of tracking the stator resistance variation very well. It is also seen that the compensator can overcome the problem of instability caused by a large mismatch between the value used in the state observer and the actual one.
8. A robust motor speed estimator based on MRAS is presented that estimated motor speed accurately for a sensorless indirect field-oriented control system. The validation of the induction motor drive was performed using both simulated and experimental implementations.
9. The proposed H_∞ controller showed a robust performance against variations of the parameters, a smaller settling time due to unit step of speed command as well as smaller maximum speed dip when the motor is subjected to step change of the load as compared with classical one.
10. The predictive motor state observer made possible the estimation of rotor flux with very low sensitivity to parameters variations and then contributed high dynamic performance compared to conventional vector control.
11. The estimated speed response with the proposed scheme gives a desired dynamic performance and a zero steady-state error, which is not affected by the load torque disturbance and the variation of the motor parameters.

6.2   Suggestion for Future Work

The presented work in this book can be expanded in several ways, including:

1. Design of adaptive speed observer for sensorless direct torque control of induction motor drives can be studied.
2. Implementation of nonlinear adaptive state observers for control the permanent magnet synchronous motor can be investigated.
3. The effects of magnetic saturation and core loss have to be taken into consideration in the machine model for studying the adaptive state observer and their compensation.
4. Design of adaptive state observers for sensorless vector control induction motor to achieve high efficiency performance.

Appendix A
Technical Specifications of the PV Module and Induction Motor Under Study

A.1 PV Module Specifications

Table A.1 summarizes the technical specification of the PV array used in the study presented in Chap. 4.

Table A.2 concludes the configuration and nominal voltage, current and power rating of the PV array used for study of this book in Chap. 4.

Table A.1 PV module technical specifications

PV module parameter	
Parameter	Values
Ns	36
Isc (A)	8.93
Voc (V)	22.66
Current at MPP, Imp (A)	8.25
Voltage at MPP, Vmp (V)	17.54
Io1 = Io2 (A)	2.04667641e−10
Rp (Ω)	129.520708
Rs (Ω)	0.30

Table A.2 PV array nominal voltage, current and power rating and configurations

PV array data	
Voltage at MPP, Vmp_array (V)	614 V
Current at MPP, Imp_array (A)	16.5 A
Power at MPP, Pmp_array (W)	10 kW
Nser (no. of modules connected in series)	35
Npar (no. of modules connected in parallels)	2

© The Author(s), under exclusive license to Springer Nature Singapore Pte Ltd. 2020
A. A. Z. Diab et al., *Development of Adaptive Speed Observers for Induction Machine System Stabilization*, SpringerBriefs in Electrical and Computer Engineering,
https://doi.org/10.1007/978-981-15-2298-7

A.2 Induction Motor Specifications

Table A.3 presents the technical specification of the induction motor used for study in Chap. 4.

Table A.3 Parameters of the induction motor

Parameters of induction motor	
Nominal power	7.5 kW
Nominal line-to-line voltage	400 V
Nominal frequency	50 Hz
Stator resistance and inductance	0.7384 Ω, 0.003045 H
Rotor resistance and inductance	0.7402 Ω, 0.003045 H
Mutual inductance	0.1241 H
Number of pole pairs	2

Appendix B
Induction Motor Parameters and Technical Specifications

Table B.1 summarizes the technical specifications of the induction motor used for experimental study in Chap. 5.

Table B.1 Parameters and data specifications of the induction motor

Parameter	Value
Rated power (W)	180
Rated voltage (V)	220
Rated current (A)	1.3–1.4
Rated frequency (Hz)	60
R_s (Ω)	11.29
R_r (Ω)	6.11
L_s (H)	0.021
L_r (H)	0.021
L_m (H)	0.29
J (kg m^2)	0.00940
Rated rotor flux (wb)	0.3
Rated speed (rpm)	1750

© The Author(s), under exclusive license to Springer Nature Singapore Pte Ltd. 2020
A. A. Z. Diab et al., *Development of Adaptive Speed Observers for Induction Machine System Stabilization*, SpringerBriefs in Electrical and Computer Engineering,
https://doi.org/10.1007/978-981-15-2298-7

Appendix C
Digital Signal Processor Kit

C.1 DSP Controller Design

Today's low-cost, high-performance DSP controllers, such as the Texas Instruments (TI) TMS320F280x, provide an improved and cost effective solution for controllers design. The F280 has integrated peripherals specifically chosen for embedded control applications. These include Analog-to-Digital converters (A/D), PWM outputs, timers, protection circuitry, serial communications, and other functions. High CPU bandwidth and the integrated power electronic peripherals of these devices make it possible to implement a complete digital controller. Most instructions for the F280, including multiplication and accumulation (MAC) as one instruction, are single cycle. Therefore, multiple control algorithms can be executed at high speed, making it possible to achieve the required high sampling rate for good dynamic response. This also makes it possible to implement multiple control loops of the inverter in a single chip and increase integration and lower system cost. Digital control also brings the advantages of programmability, immunity to noise, and eliminates redundant voltage and current sensors for each controller. With fewer components, the system requires less engineering time, and it can be made smaller and more reliable. The extra DSP bandwidth is available for implementing more sophisticated algorithms, as well as communications to host systems and I/O devices such as LCD displays. DSP programmability means that it is easy to update systems with enhanced algorithms for improved reliability [1].

C.2 DSP Based High Voltage Digital Motor Control Kit

The High Voltage Digital Motor Control (DMC) and Power Factor Correction (PFC) kit (TMDSHVMTRPFCKIT), provide an easy way to learn and experiment with digital control of high voltage motors [2] (Figs. C.1 and C.2).

© The Author(s), under exclusive license to Springer Nature Singapore Pte Ltd. 2020   77
A. A. Z. Diab et al., *Development of Adaptive Speed Observers for Induction Machine System Stabilization*, SpringerBriefs in Electrical and Computer Engineering, https://doi.org/10.1007/978-981-15-2298-7

Fig. C.1 Photo of Texas Instruments High Voltage Motor Control and PFC Developer's kit
TMDSHVMTRPFCKIT

C.2.1 Features of the TMS320F28035 System

Table C.1 gives a list of the TI DSP chip parameters and their corresponding values.

The High Voltage Digital Motor Control Kit is equipped with the following list
of peripherals:

- F28035 control card.
- High Voltage DMC board with slot for the control card.
- 15 V DC Power Supply.
- AC power Cord, Banana Plug Cord, USB Cable.
- CCS4 CD & USB Stick with Quick Start GUI and Guide.

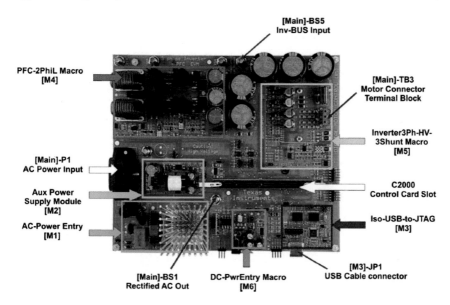

Fig. C.2 TMDSHVMTRPFCKIT board structure

Table C.1 TI chip parameters and their corresponding values

Parameter name	TMS320F28035
MIPS	120
Frequency (MHz)	60
RAM (KB)	20 KB
Flash (KB)	128 K
Boot loader available	Flash
External memory interface	Yes
PWM channels	14
12 bit A-D (number of channels)	16
Conversion time (μs)	6.1
PWM timer comparators	2
Total serial ports	2

C.2.2 Features of the High Voltage Motor Control and PFC Board

- 3-Phase Inverter Stage capable of sensorless and sensored field-oriented control (FOC) of high voltage ACI and PMSM motor and trapezoidal, and sinusoidal control of the high voltage BLDC motor. 350 V DC max input voltage and 1 kW maximum load in the configuration shipped.

- Power Factor Correction stage rated for 750 W. Takes rectified AC input (85–132VAC/170–250VAC). 400 V DC Max output voltage.
- AC Rectifier stage rated for 750 W* power. Accepts 85–132VAC/170–250VAC input.
- Aux Power Supply Module (400 V to 15 V & 5 V module) generates 15 and 5 V DC from rectified AC voltage or the PFC output (input Max voltage 400 V, min voltage 90 V).
- Isolated CAN, SCI & JTAG.
- Four PWM DACs to observe the system variables on an oscilloscope.
- Hardware Developer's Package available which includes schematics and bill of materials.
- Open source software available through Code Composer Studio (CCS) and ControlSUITE for each type of motor and control systems.

References

1. http://www.ti.com/product/TMS320F28035. Accessed April 2019
2. http://www.ti.com/lit/ml/sprugu7/sprugu7.pdf. Accessed April 2019

Printed in the United States
By Bookmasters